INTRODUCTION TO ROBOTICS

A Systems Approach

JAMES A. REHG

Director, Robotics Resource Center

*Piedmont Technical College
Greenwood, South Carolina*

PRENTICE-HALL, INC.
Englewood Cliffs, New Jersey 07632

*To the three women
who most influenced my life . . .*

*Rose
Marci
Bettie*

CONTENTS

PREFACE ix

CHAPTER 1 INTRODUCTION TO INDUSTRIAL ROBOTS 1

    1-1 Introduction 3
    1-2 History of the Industry 3
    1-3 A 20-Year Old Industry 5
    1-4 Definition of Robotics 7
    1-5 The Robot System 8
           The basic system 8 Mechanical arm 8 Production tooling 10
           External power source 10 Robot controller 13
           Teach stations 14
    1-6 Some Basic Terms 17
    1-7 Summary 20
          Questions 20

CHAPTER 2 ROBOT CLASSIFICATION 23

    2-1 Why Classify? 25
    2-2 Robot Arm Geometry 25
           Work envelope analysis by coordinate systems 25
           Rectilinear coordinate systems 27 Cylindrical coordinate systems 29 Spherical coordinate robots 34
           Jointed-spherical coordinate systems 36
    2-3 Power Sources 39
           Hydraulic drive 39 Pneumatic drive 41
           Electric drive 41
    2-4 Application Areas 43
           Assembly 43 Non-assembly 45
    2-5 Control Techniques 45
           Closed-loop system 46 Open-loop system 47
    2-6 Path Control 49
           Stop-to-stop 51 Point-to-point 51 Controlled path 55
           Continuous path 56

Contents   v

      2-7   Controller Intelligence   57
      2-8   Summary   57
            Questions   58

CHAPTER 3   END-OF-ARM TOOLING   61

      3-1   General Characteristics   63
      3-2   Conservation of Job IQ   63
      3-3   Classification   65
            Standard grippers 65   Vacuum grippers 68
            Vacuum surfaces 71   Vacuum suckers 72
            Magnetic grippers 72   Air pressure grippers 72
            Special purpose grippers 76
      3-4   Special-Purpose Tools   77
      3-5   Assembly Fixtures   80
            Compliance 81
      3-6   Multiple End Effector Systems   85
      3-7   Vision and Artificial Skin   87
      3-8   Summary   87
            Questions   88
            Projects   89

CHAPTER 4   ROBOT AND CONTROLLER OPERATION   91

      4-1   Introduction   93
      4-2   Reference Frames   94
      4-3   Open-Loop Systems   96
            Internal control 97   External control 98
            Operation 100
      4-4   Closed-Loop Systems   100
            Internal control 101   Potentiometers 101
            Optical encoders 103   Resolvers and synchros 106
            External control 110   Operation 111
      4-5   Controller Architecture   111
      4-6   Summary   115
            Questions   116

CHAPTER 5   SENSORS AND INTERFACING   119

      5-1   Introduction   121
      5-2   Contact Sensors   122
            Limit switches 122   Artificial skin 127

5-3 Non-Contact Sensors 129
*Proximity sensors 130 Photoelectric sensors 135
Vision systems 142 Vision system components 143
Image measurement 145 Image analysis 145
Image recognition 147*
5-4 Process Sensors 149
5-5 Interfaces 151
*Simple sensor interface 152 Wrist interface 154
Robot control interface 155 Joint control level 155
Coordinate transform level 156 Trajectory control
level 158 Complex sensor interface 158*
5-6 Summary 162
Questions 162

CHAPTER 6 ROBOT PROGRAMMING 165

6-1 Introduction 167
6-2 Robot Language Development 167
6-3 Language Classification 168
*Joint control languages 169 Primitive motion
languages 170 Structural programming languages 171
Task-oriented languages 172*
6-4 Sample Programs 173
6-5 T3 Language 173
*T3 procedure 173 T3 commands 176 Input and output
signals 177 Sequence functions 178*
6-6 Program Analysis 179
6-7 Summary 183
Questions 184
Projects 184

CHAPTER 7 SAFETY 185

7-1 Introduction 187
7-2 OSHA 187
7-3 General Personnel Safety 188
7-4 Operator and Maintenance Personnel Safety 189
7-5 Summary 190
Questions 190

CHAPTER 8 HUMAN INTERFACE: OPERATOR TRAINING, ACCEPTANCE, PROBLEMS 191

8-1 Introduction 193
8-2 General Training 193
   General training programs 194
8-3 Operator Training 195
8-4 Maintenance Training 195
8-5 Resistance 196
8-6 Organized Labor 197
8-7 Summary 197
   Questions 197

APPENDIX A 199

APPENDIX B 206

INDEX 227

BIOGRAPHY

James A. Rehg, CMfgE, is director of the Robotics Resource Center and head of the Automated Manufacturing Technology department at Piedmont Technical College. He earned both a B.S. and M.S. in Electrical Engineering from St. Louis University and has completed additional graduate study at Wentworth Institute, University of Missouri, South Dakota School of Mines and Technology, and Clemson University. Before coming to Piedmont Technical College he was department head of Electronic Engineering Technology at Forest Park Community College in St. Louis, Missouri and a Senior Instrumentation Engineer for McDonnell Douglas Corporation in St. Louis.

Professor Rehg has authored and presented numerous papers on subjects directly related to training in automation and robotics. He has also been a consultant to nationally-recognized corporations and many educational institutions. He has led numerous seminars and workshops in the areas of robotics and microprocessors and has developed extensive seminar training material.

As a Certified Manufacturing Engineer in robotics, and as a member of the Education and Training Division of Robotics International of SME, he has received national recognition for his work in robotics education.

PREFACE

An industrial robot is just another industrial machine. This statement has frequently been used by those in education and industry to calm fears of mass worker displacement or to encourage corporate management to make the jump into flexible automation. Like the NC turning center or the lathe which preceded it, the industrial robot is a machine designed to aid in the production process, but the robot is *not* just another industrial machine.

Robots can be used in every industry providing goods and services, can be adapted to numerous job functions, can easily change job functions, and will work with uncanny skill and unmatched endurance. Robots are different from any industrial machine in the history of automated production. A robot is part of a production system which includes the production machinery, the robot, sensors, safety hardware, and a computer control to orchestrate the entire operation.

Education in robotics must have a systems emphasis for an understanding of the total production processes present in industry. This book addresses robotics at the systems level. In addition, computer- and microprocessor-driven automation is changing at a rapid pace with sophistication increasing with every new machine introduction and software update. The systems material in the text views the current technology but also looks into the future of flexible automation with robots to describe and explain the future direction of robotics systems.

The book is written both for the industrial reader who wants an introduction to flexible automation systems using robotics and also for the student in two- and four-year colleges. To support classroom instruction, questions and projects are included at the end of each chapter. The text can be used for a one quarter or one semester course to introduce robotics or flexible automation systems. **Chapter 1** introduces industrial robots and the work cell system in which they operate. The definition of an industrial robot and the new terms used to describe its operation are also included. In **Chapter 2,** the different types of robot systems are classified by arm geometry, power sources, applications, control techniques, path control and controller intelligence. The advantages and disadvantages of each type of system are discussed.

The end-of-arm tooling used on current and future robot systems is described in **Chapter 3.** The material includes standard tooling and special purpose grippers used on current applications, along with a discussion of artificial skin, which will be used on future tactile sensing machines. In **Chapter 4,** the operation of robot controllers is described with an emphasis on open- and closed-loop systems, and the positioning devices that make them function. **Chapter 5** includes contact, non-contact, and process sensors used in automatic work cells and the technology required to interface them at the machine and systems level. An overview of robot programming is included in **Chapter 6** with the current languages classified into four levels. A sample program to solve a typical manufacturing problem using the Cincinnati Milacron T3 language is provided. In **Chapter 7,** the safety of both humans and machines in a robot work cell is discussed, and in **Chapter 8,** the problems created by automation within the human interface are covered.

Chapters 2 through 8 are independent of each other and can be read or used in any order. In a semester system, Chapters 1 through 7 would support a good introductory course in robotics, but if the course is taught in the quarter time frame, Chapters 5 and 6 may have to be included in a later course. An instructor's manual, available with a course outline for an introductory course in robotics, includes answers to all end-of-chapter questions and projects.

Many people have contributed to the preparation of this manuscript. I would like to thank the many companies that have contributed material and have been supportive and helpful. A very special thanks to Richard Upchurch for many enlightening discussions, Marci Rehg and Bettie Steffan for suggestions and editing throughout the text, and the students in Automated Manufacturing Technology at Piedmont Technical College who used the rough draft material and offered many useful suggestions.

JAMES REHG

1

INTRODUCTION TO INDUSTRIAL ROBOTS

The fact is, that civilization requires slaves. The Greeks were right there. Unless there are slaves to do the ugly, horrible, uninteresting work, culture and contemplation become almost impossible. Human slavery is wrong, insecure and demoralizing. On mechanical slavery, on the slavery of the machine, the future of the world depends.

OSCAR WILDE

1-1 INTRODUCTION

The introduction of new technology in the industrial nations precipitates changes throughout the social structure of every country in the world. The nature and degree of those changes are proportional to the effects a new technology has on the production of goods and services. Technology in agriculture, for example, caused the food production industry to change from an employer of 80 percent of the population in 1890 to an employer of 3 percent of the population in 1983. There was a reduction in the number of farmers but an increase in the number of other jobs required to support the automated farm industry. No one can predict the effect that robotics will have as it emerges from its 20-year infancy to an industrial force with greater change potential than that of any industrial machine in the past. Will robot technology produce more jobs than it displaces? What types of jobs will be required with this new technology? What part will the robot play in the "factory of the future"? These questions and many more need to be answered before the full impact of robots can be understood. The first step in the process of understanding the impact of robots is to understand the machines themselves. What are industrial robots, what can they do, and what are they currently not able to do? The answers to these questions are provided in the following chapters.

1-2 HISTORY OF THE INDUSTRY

A brief review of robot development is important because it puts the current machines and interest in them into a historical perspective. The following list of dates highlights the growth of automated machines which led to the development of the industrial robots currently available.

- **1801** Joseph Jacquard invents a textile machine which is operated by punch cards. The machine is called a programmable loom and goes into mass production.
- **1830** American Christopher Spencer designs a cam-operated lathe.
- **1892** In the United States, Seward Babbitt designs a motorized crane with gripper to remove ingots from a furnace.
- **1921** The first reference to the word *robot* appears in a play opening in London. The play, written by Czechoslovakian Karel Capek, introduces the word *robot* from the Czech *robota*, which means a serf or one in subservient labor. From this beginning the concept of a robot takes hold.
- **1938** Americans Willard Pollard and Harold Roselund design a programmable paint-spraying mechanism for the DeVilbiss Company.

1946 George Devol patents a general purpose playback device for controlling machines. The device uses a magnetic process recorder.

In the same year the computer emerges for the first time. American scientists J. Presper Eckert and John Mauchly build the first large electronic computer called the Eniac at the University of Pennsylvania. A second computer, the first general-purpose digital computer, dubbed Whirlwind, solves its first problem at M.I.T.

1948 Norbert Wiener, a professor at M.I.T., publishes *Cybernetics*, a book which describes the concept of communications and control in electronic, mechanical, and biological systems.

1951 A teleoperator-equipped articulated arm is designed by Raymond Goertz for the Atomic Energy Commission.

1954 The first programmable robot is designed by George Devol, who coins the term Universal Automation. He later shortens this to Unimation, which becomes the name of the first robot company.

1959 Planet Corporation markets the first commercially available robot.

1960 Unimation is purchased by Condec Corporation and development of Unimate Robot Systems begins.

American Machine and Foundry, later known as AMF Corporation, markets a robot, called the Versatran, designed by Harry Johnson and Veljko Milenkovic.

1962 General Motors installs the first industrial robot on a production line. The robot selected is a Unimate.

1964 Artificial intelligence research laboratories are opened at M.I.T., Stanford Research Institute (SRI), Stanford University, and the University of Edinburgh.

1968 SRI builds and tests a mobile robot with vision capability, called Shakey.

1970 At Stanford University a robot arm is developed which becomes a standard for research projects. The arm is electrically powered and becomes known as the Stanford Arm.

1973 The first commercially available minicomputer-controlled industrial robot is developed by Richard Hohn for Cincinnati Milacron Corporation. The robot is called the T3, The Tomorrow Tool.

1974 Professor Scheinman, the developer of the Stanford Arm, forms Vicarm Inc. to market a version of the arm for industrial applications. The new arm is controlled by a minicomputer.

1976 Robot arms are used on Viking 1 and 2 space probes.

Vicarm Inc. incorporates a microcomputer into the Vicarm design.

1977 ASEA, a European robot company, offers two sizes of electric-powered industrial robots. Both robots use a microcomputer controller for programming and operation. In the same year Unimation purchases Vicarm Inc.

1978   The Puma (Programmable Universal Machine for Assembly) robot is developed by Unimation from Vicarm techniques and with support from General Motors.

1980   The robot industry starts its rapid growth, with a new robot or company entering the market every month.

1-3 A 20-YEAR-OLD INDUSTRY

A review of history reveals that robots are not new and that the application of robots to solve industrial problems also dates back to the early 1960s. Another surprise is that the mechanical components of robot-like devices date back to the early part of this century. With robots and robot technology available in 1960, why did the United States robot movement wait until the late 1970s to get started? Two reasons explain the rapid development in robotics during the last decade.

The first reason relates to *economics* and the second to *hardware*. A number of barriers prohibit a rapid deployment of robots into industrial applications. The first and primary barrier is economics. The application of a robot must be justifiable from an economic standpoint, which means that the machine must pay for itself in one to two years. The primary cost savings used to calculate payback in industry are direct and indirect labor costs. This means that the cost of employing a human operator, who will be replaced in the work cell by the robot, is the primary savings value which results from the purchase of a robot. The operator cost thus includes the direct labor cost of base salary plus the cost of all fringe benefits such as vacation, sick leave, and retirement, which are lumped under indirect labor cost. In 1960 the cost of operating a robot was over nine dollars per hour, while the overall cost of a human operator was less than five dollars per hour. When these figures are compared to equivalent data from 1982, the present advantage of flexible automation in the form of robots becomes apparent. The hourly labor cost in the automotive industry, for example, has climbed to over 20 dollars per hour while the hourly cost to operate a robot has dropped to less than six dollars. The average hourly cost in the United States for production of all durable goods was about 15 dollars. These values are illustrated in Figure 1-1. The primary cause of the separation in the wage scales of human operators and robots was the *inflation* of the 1970s. Although a good human operator will outperform a robot in most applications, the present shortage of skilled operators and the increase in wages, primarily from inflation, has given the robot industry in America an edge in the market since the late 1970s.

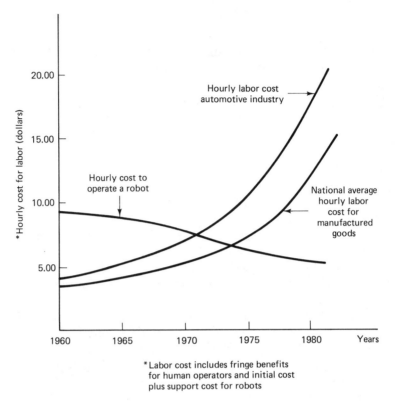

Figure 1-1 Robot and Human Labor Cost

The second factor which influenced the emergence of the robot industry in the United States after nearly 20 years of flat growth was the development of the *microprocessor*, or computer on a chip. The robots available in 1970 had remarkable capability, but their versatility pales before that of machines currently being offered by robot manufacturers. The primary difference is not mechanical but electronic sophistication. The microprocessor which was developed by Intel Corporation in 1971 has provided robot manufacturers with a source of very inexpensive intelligence to drive their machines. One of the first computer-controlled robots was developed by Cincinnati Milacron in the early 1970s. The machine used a single minicomputer to direct all the robot motion and machine control within the work cell. The machine had tremendous capability but was limited by the speed of its single computer, which had to do all the calculations. Thus the development of the microprocessor permitted companies like Cincinnati Milacron to incorporate many computers into the controller which drives the robot arm. Their new controller has a family of microcomputers which operate simultaneously to solve the many concurrent problems encountered with a robot in motion. The added intelligence and operating speed along with improved software in the current robot system have opened up many new applications not formerly possible. Of course, the explosive growth of the microelectronic industry has spawned equivalent growth in many related industries. For robotics, however, the development of the microprocessor was crucial and the reason robots emerged from 20 years of flat growth.

1-4 DEFINITION OF ROBOTICS

To define a robot in a way that is generally acceptable to every manufacturer and user is difficult. However, the importance of a clear definition becomes apparent when the numbers of robots in use in various countries and industries are counted and reported. In addition, many single-purpose machines, often called hard-automation, have some features which make them look like robots. Without some definition, the number of robots reportedly in use in Japan would total over 85,000. If the definition developed by the Society of Manufacturing Engineers (SME) is applied to the 85,000 machines, only approximately 12,000 would qualify as robots. The definition developed by the Robot Institute of America, a group within SME composed of robot-equipment manufacturers, is as follows:

> *A robot is a reprogrammable multifunctional manipulator designed to move material, parts, tools, or specialized devices through variable programmed motions for the performance of a variety of tasks.*

The key words are *reprogrammable* and *multifunctional*, since most single-purpose machines do not meet these two requirements. Reprogrammable means that the machine must be capable of being reprogrammed to perform a new or different task or to be able to change significantly the motion of the arm or tooling. Multifunctional emphasizes the fact that a robot must be able to perform many different functions, depending on the program and tooling currently in use. A variation on this definition which more clearly describes the intelligence of current robots is as follows:

> *A robot is a one-armed, blind idiot with limited memory which cannot speak, see, or hear.*

Despite the tremendous capability of currently available robots, even the most poorly prepared worker is better equipped than a robot to handle many of the situations which occur in the work cell. Workers, for example, realize when they have dropped a part on the floor or when a parts feeder is empty. Without a host of sensors a robot simply does not have any of this information; and even with the most sophisticated sensor system available, a robot cannot match an experienced operator. The design of a good automated work cell therefore requires the use of peripheral equipment interfaced to the robot controller to even approximate the sensory capability of a human operator.

1-5 THE ROBOT SYSTEM

At present, the difference between the robot system and the work cell system is quite clear. The robot system includes only the robot hardware, whereas the work cell system includes the robot system components plus all the additional equipment required to produce the product. As manufacturers start integrating robot automation into increasingly complex production systems, the distinction between the robot system and the production system becomes less clear. For example, in many future work cells the robot will be programmed and controlled by a computer-aided design and manufacturing system (CAD/CAM). In others, the tooling on the robot arm will be changed automatically by the system, depending on the needs of the production process. A study of the basic robot system is the logical place to begin in order to understand the current state-of-the-art and possible future developments.

The Basic System

A basic robot system is illustrated in the block diagram in Figure 1-2. The system includes a mechanical arm to which the end-of-arm tooling is mounted, a computer-based controller with attached teach station and storage device, and a source for pneumatic or/and hydraulic power.

Mechanical Arm

The arm is a mechanical device driven by either electric servo-motors, pneumatic devices, or hydraulic actuators. The basic drive elements will be either linear or rotary actuators, integrated into the mechanical arm to provide either linear arm motion or rotational motion. The combination of motions included in the arm will determine the type of arm geometry which is present. The basic geometries include rectangular, cylindrical, spherical, and jointed-spherical. Figure 1-3 shows a jointed-spherical mechanical arm with all the motions indicated. Note that all six motions are rotational, and that five of the six are produced by rotational actuators. The elbow extension is the only rotary motion produced by a linear actuator.

The six motions are divided into two groups. The first group includes

Sec. 1-5   The Robot System

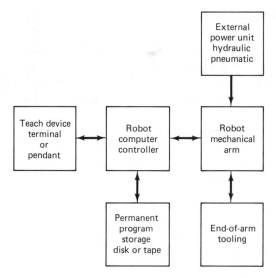

Figure 1-2 Basic Robot System

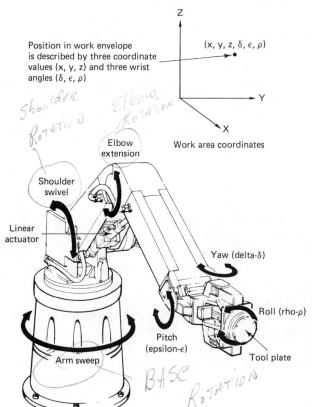

Figure 1-3 Jointed-Spherical Mechanical Arm (*Courtesy of Cincinnati Milacron*)

the arm sweep, shoulder swivel, and elbow extension and is called the *position motions*. With a combination of these three motions the arm can move to any required position within the work area. The second group of motions includes the pitch, yaw, and roll associated with the wrist which is located at the end of the robot arm. A combination of these three motions, called *orientation*, permits the wrist to orient the tool plate and tool with respect to the work.

A picture of the Cincinnati Milacron T3-566 arm appears in Figure 1-4. A complete description of the arm geometries and drive systems is presented in Chapter 2.

Production Tooling

The robot arm alone has no production capability, but the robot arm interfaced to production tooling becomes an effective production system. The tooling to perform the work task is attached to the tool plate at the end of the arm. The tool plate, which is usually a part of the wrist, is identified in Figure 1-3.

The tooling is frequently identified by several names. The term used to describe tooling in general is "end-of-arm tooling" or "end effector." If the tooling is an open-and-close mechanism to grasp parts, it is referred to as a "gripper." In this text the terms *gripper* and *end-of-arm tooling* will both be used to refer to the tool on the robot arm.

Figure 1-5 is a picture of a Fibro Manta gripper attached to the tool plate of a Cincinnati Milacron T3-566 arm by an adapter plate. A complete discussion of robot end-of-arm tooling is presented in Chapter 3.

External Power Source

The external power sources required to operate a robot system include the electricity to power the electronic controller and a hydraulic or pneumatic source to operate the arm motion and end-of-arm tooling. Since most grippers are activated by compressed air and some arms use pneumatic devices to provide the motion, a source of compressed air is required for most systems. Large robot arms using hydraulic actuators for motion require a hydraulic power source, and in some cases a compressed air source will also be necessary for the gripper. All electric drive arms require only electrical power for motion, but many will need compressed air for tooling.

The hydraulic T3-566 robot uses the pump and tank assembly in Figure 1-6 to generate the 2300 psi fluid pressure necessary to operate the arm. In addition, there is a power panel for system power control.

Sec. 1-5   The Robot System

Figure 1-4 Cincinnati Milacron T3-566 Mechanical Arm

Figure 1-5 Fibro Manta Gripper

Figure 1-6 Hydraulic Power Unit for the T3-566 Cincinnati Milacron Robot

Sec. 1-5   The Robot System

Robot Controller

Of all the building blocks of a robot system, the controller is the most complex. It is also the unit with the greatest degree of variation from one manufacturer to the next. Figure 1-7 is a basic block diagram of the new controller used on all Cincinnati Milacron electric robots, and Figure 1-8 is a picture of the actual unit.

The controller, basically a special-purpose computer, has all the elements commonly found in computers, such as a central processing unit (CPU), memory, and input and output devices. This particular controller has a network of CPUs, each having a different responsibility within the system. The entire network of CPUs, known as the controller computer, has the primary responsibility for controlling the robot arm and the work cell in

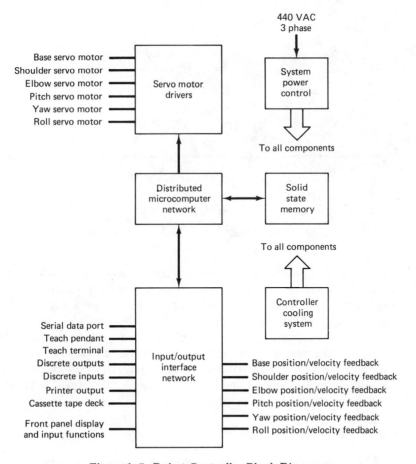

Figure 1-7 Robot Controller Block Diagram

Figure 1-8 Version 4.0 Controller for T3-726 Cincinnati Milacron Robot

which it is operating. The controller receives feedback from the arm on joint position and velocity, and responds with outputs to the servo drives to change the current position or velocity based on the program stored in memory. The controller can also communicate with other computer controlled machines, the teach stations, a permanent storage device (see Figure 1-9), and discrete devices in the work cell. The front panel display and system power control module are a part of every controller. In addition, many controllers are cooled internally by air conditioners or fans. The operation of a controller is discussed in detail in Chapter 4.

Teach Stations

Teach stations on currently available robots may consist of teach pendants, teach terminals (see Figure 1-10), or controller front panels (see Figure

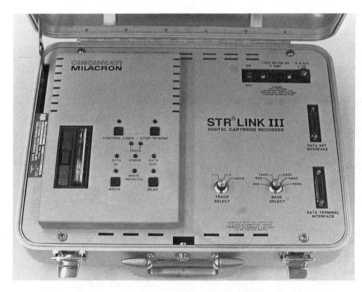

Figure 1-9 Permanent Program Storage

Figure 1-10 Teach Terminal and Pendant for Cincinnati Milacron T3-726 System

1-11). Some robots permit a combination of these units to be used in programming the robot system, while others provide only one system programming unit. Teach stations accomplish three objectives: (1) they *power up the robot and prepare it to be programmed*; (2) they *write and teach programs* that result in the solution of manufacturing problems; and (3) they *execute programs* in the production work cell. The development of a program includes keying in commands in the controller language, the physical movement of the robot arm to the desired position in the work cell, and the recording of that position in memory. Chapter 6 includes a detailed analysis of the programming process.

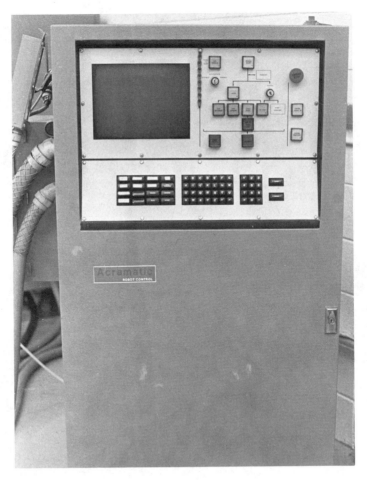

Figure 1-11 Front Panel Cincinnati Milacron Version 3.0 Controller

1-6 SOME BASIC TERMS

With the introduction of every new technology there follows a new vocabulary which must be learned in order to be literate in that field. Robotics is no exception. Much of a robotics vocabulary can be learned as required, but a few terms are prerequisites before additional material can be learned.

Accuracy. Accuracy is the degree to which a robot arm is able to move to a specific point in the work cell when the point coordinates are entered from an off-line programming station. For example, a remote terminal could specify that the arm should move to a point described as $X = 50.00$ in., $Y = 45.300$ in., $Z = 10.01$ in., delta $= 0.00$ deg., epsilon $= 90.00$ deg., and rho $= 100.00$ deg. The accuracy specification would describe how close the robot arm approximates the desired point.

Repeatability. Repeatability is the degree to which a robot system is able to return to a specific point in the work cell which was taught using the teach pendant. In this procedure a robot is taught the required point by moving the arm to the location with the teach pendant and then pushing the program point button. The repeatability specification thus describes how closely the robot arm reproduces the taught point on each cycle of the program execution.

Work envelope. This term refers to the volume of space in which the arm can move. Any parts or equipment within this volume can be reached by the robot gripper. Figure 1-12 shows a work envelope for a spherical geometry robot.

Degree of freedom. Every joint or movable axis on the arm is considered a degree of freedom. A machine with six movable joints, such as the one in Figure 1-3, is a robot with six degrees of freedom. *Orientation* of the tool by the wrist involves a maximum of three degrees of freedom, and up to four degrees of freedom are used for *positioning* within the work envelope.

Orientation axes. The orientation axes in Figure 1-3 are labeled *pitch*, *yaw*, and *roll*. These three axes constitute the basic wrist movements for all robot geometries. *Pitch* is rotation about the transverse axis (movement in the vertical plane), *yaw* is rotation about the perpendicular axis (movement in the horizontal plane), and *roll* is rotation about the longitudinal axis.

Position axes. The position axes in Figure 1-3 are labeled *arm sweep*, *shoulder swivel*, and *elbow extension*. A fourth position axis, *linear movement* of the entire robot along the floor, is not shown in the figure. The

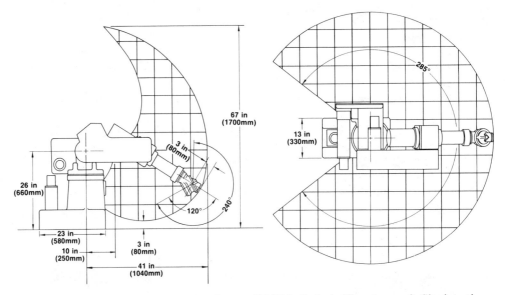

Figure 1-12 Work Envelope for a T3-726 Robot (*Courtesy of Cincinnati Milacron*)

type of motion for each degree of freedom is dependent upon the arm geometry of the robot. For example, the jointed-spherical machine in Figure 1-3 has different motions along the position axes from those of a cylindrical machine.

Tool-center-point. The *point of action* for the tool mounted to the robot tool plate is called the tool-center-point. With a tool-center-point of 0, 0, 0 for L, A, and B, respectively, the tool-center-point is located at point TCP1 in Figure 1-13. With L = 10 in., A = 5 in., and B = 4 in. the tool-

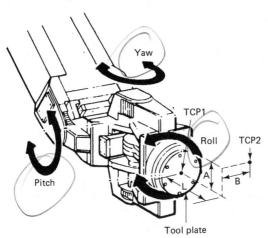

Figure 1-13 Tool Center Point (*Courtesy of Cincinnati Milacron*)

Sec. 1-6   Some Basic Terms   19

center-point is located at TCP2. When a tool is mounted to the tool plate, the distance from the TCP1 to the point on the tool where the work is contacted is determined and entered into the controller through variables L, A, and B. For example, the welding torch in Figure 1-14 has a TCP of L = 14 in., A = −4 in., and B = 0 in. The robot will control the motion at the tool-center-point as the arm moves through the programmed points.

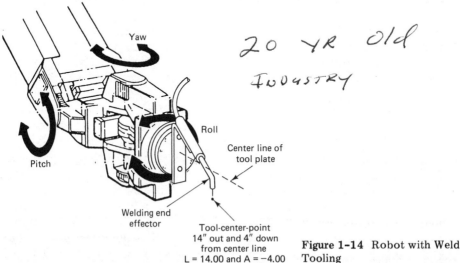

Figure 1-14 Robot with Welding Tooling

Work cell coordinates.   All points programmed in the work cell are identified by the *X, Y, and Z coordinate values* of the tool-center-point, along with the *joint angles* of the wrist axes, δ (delta), ϵ (epsilon), and ρ (rho). The coordinates put the tool-center-point in the correct position, and the wrist joint angles adjust pitch, yaw, and roll for the proper tool orientation. Figure 1-3 shows the coordinate system frequently used by robot manufacturers.

Speed.   The rate at which the robot can move the tool-center-point under program control is a measure of the machine's speed. Speed is expressed in inches per second or millimeters per second, and the maximum speed is an important robot specification.

Payload.   The payload or lifting capacity of a robot often determines if the machine is suited for a specific task. The robot payload is quoted in pounds or kilograms at a specified distance from the tool plate, and it is the maximum weight that the machine can move under program control and still maintain the repeatability and reliability specifications. The maximum payload is the total combined weight of the gripper or end-of-arm tooling and the part to be moved.

Additional terms associated with the robot hardware and software are introduced as needed throughout the remainder of the text.

1-7 SUMMARY

Automation is going to change the work place, but the jury is still out on how it will affect workers and their on-the-job responsibilities. A study of the history of robotics indicates that the technology is not new but almost 20 years old. The recent heightened interest is a result of two factors, one *economic* and one *technical*. *High inflation* during the 1970s, which drove labor costs up, led to increased applications with the less labor-intensive robot technology. The technical factor in the emergence of the robot industry is the development of the *microprocessor*. Robot designs could take advantage of the low-cost computer power provided by the explosive growth of the microprocessor and microcomputer industry. More intelligent robots with increased software capability could solve more of the problems present in manufacturing.

The definition of robot hardware includes two very important terms: *reprogrammable* and *multifunctional*. In order to qualify as a robot the machine's hardware must be capable of changing its functional characteristics by changing the program which drives it through its motions. The basic robot system includes a mechanical arm, special tooling attached to the arm at the tool plate, one or more computers in a controller, a teach station, a program storage device, and a source of pneumatic or hydraulic power.

QUESTIONS

1. Identify an early robot-like device which demonstrates the mechanical operation found in later industrial robots.
2. Who was awarded the first patent for an industrial robot?
3. What were the names of the first three robot manufacturers?
4. What company was the first to drive a robot with a powerful minicomputer?
5. What two events in the 1970s caused industrial robots to emerge from 20 years of flat growth?
6. Describe how the events identified in question 5 affected the industrial robot industry.
7. What is the SME definition of a robot?
8. What operational feature makes robots different from other hard-automation machinery?
9. What are the major elements of a robot system?

10. Describe the two basic groups into which all robot motion is divided.
11. What are the three names used to identify robot tooling?
12. What function does the external power source serve in the basic robot system?
13. Describe the basic elements found in robot controllers.
14. What are the three basic operations performed from robot teach stations?
15. Define repeatability and accuracy with respect to robot operation and explain how they are different.
16. Define the following robot terms: work envelope, degrees of freedom, orientation axes, position axes, tool-center-point, maximum operational weight, speed, and work cell coordinates.

— HUMAN COSTS EXCEEDED COSTS OF ROBOT

2

ROBOT CLASSIFICATION

2-1 WHY CLASSIFY?

Today's world tends to put everything into categories, groups, and classifications. This same fate has befallen the robot industry. In this chapter the six machine classification groups used by the robot industry are explained so that a working knowledge of current robots can be developed. This chapter also provides an opportunity to compare and contrast the primary system components which make each vendor's hardware unique.

No consensus exists among robot manufacturers and users for the best classification method. One classification technique, called the *robot arm geometry* classification, divides robots into categories based on the shape of the work area produced by the robot arm. According to this technique, robots can be *rectangular, cylindrical,* or *spherical,* with subgroups within the spherical classification. Another technique measures the relative intelligence of the robot system and divides robots into *low-, medium-,* and *high-*technology machines. The remaining four classification groups included in this chapter are *power sources, applications, control techniques,* and *path control.*

2-2 ROBOT ARM GEOMETRY

In general, the three basic robot work areas, called work envelopes, are *rectangular, cylindrical,* and *spherical.* These envelopes describe the shape of the space in which an industrial robot can function. A robot with a rectangular work envelope will be able to move its gripper to any position within the cube or rectangle defined as its working volume. Figure 2-1, for example, shows the required rectangular work envelope of a robot used to load a conveyer from supply bins. A cylindrical work area robot can move its gripper within a volume that is described by a cylinder, and a spherical work area robot can move its gripper within a volume that is described by half of a sphere. A cylindrical work volume is illustrated in Figures 2-2a and 2-2b. A production cell with a front-loading furnace, parts feeder, and press would generally require a cylindrical work envelope, but under certain conditions a spherical work volume would satisfy the requirements.

Work Envelope Analysis by Coordinate Systems

A more detailed understanding of each of the work envelopes can be obtained by analyzing the various coordinate systems which produce the work envelopes. The three systems include the following types:

1. The *rectilinear coordinate system*, used with the rectangular work envelope
2. The *cylindrical coordinate system*, used with the cylindrical work envelope
3. The *spherical coordinate system*, used with the spherical work envelope

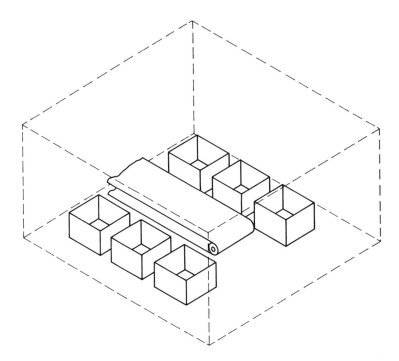

Figure 2-1 Rectangular Work Volume

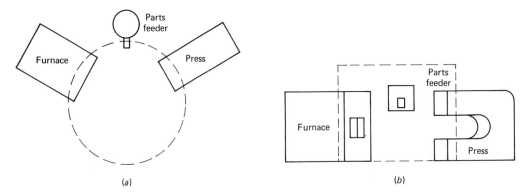

Figure 2-2 Cylindrical Work Volume

Rectilinear Coordinate Systems

A robot with rectilinear geometry is illustrated in Figure 2-3. In the illustration, the three degrees of freedom for positioning are indicated by arrows to show movement in the X, Y, and Z directions; the three degrees of freedom for orientation of the tool mounting plate on the wrist are also shown by arrows. The orientation of the tool is accomplished with rotation about axes δ, ϵ, and ρ which provide the pitch, yaw, and roll motion always associated with wrist movement. Movement in one of the degrees of freedom for positioning requires only one axis to change. For example, to move the wrist vertically only the Z axis must change. However, on the other robot geometries two or more axes must change for movement along a coordinate.

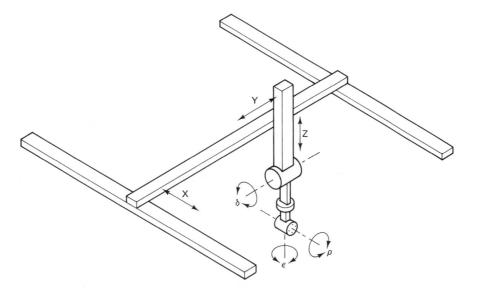

Figure 2-3 Rectilinear Geometry Robot

The power for movement in the X, Y, and Z directions is provided by linear actuators or by gear-driven mechanisms. However, rotary actuators are used to provide the pitch, roll, and yaw turning motion at the tool mounting plate; these are powered by hydraulic, pneumatic, or electric sources. The rectilinear robot system must be positioned over the work area either by suspension from the ceiling or by floor supports. An example of a rectilinear type of robot can be found in Figure 2-4.

Rectilinear coordinate robot geometry has the following advantages:

- Very large work envelopes are possible since travel along the X direction can be increased easily. Systems have been developed with work envelopes over 80 feet long.

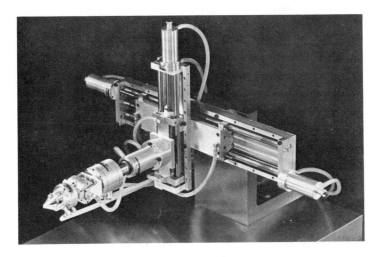

Figure 2-4 B-A-S-E Robot System from Mack Corporation (*Courtesy of Mack Corporation*)

Figure 2-5 Gantry Type Robot (*Courtesy of GCA Corporation*)

Sec. 2-2   Robot Arm Geometry

- Overhead mounting leaves free large areas of manufacturing floor space for other uses.
- Simpler control systems can be used.

The disadvantages of this type of robot geometry include the following factors:

- Access to the work envelope by overhead crane or other material-handling equipment may be impaired by the robot-supporting structure.
- On some models the location of drive mechanisms and electrical control equipment overhead makes maintenance more difficult.

The primary applications for rectilinear coordinate systems are in materials handling, parts handling related to machine loading and unloading, assembly of small systems, and in electronic printed circuit board assembly. The robot in Figure 2-5 is a gantry type rectilinear coordinate machine; the large size of the work area is evident.

Cylindrical Coordinate Systems

The six degrees of freedom for a cylindrical coordinate robot are illustrated in Figure 2-6. The wrist is positioned in the work area by two linear movements (Z and R) and one angular rotation (θ). A change in the vertical position can be achieved by a change in just one axis, resulting in vertical movement along the center member. Figure 2-7a illustrates this single-axis movement. To change the position of the gripper with respect to the center

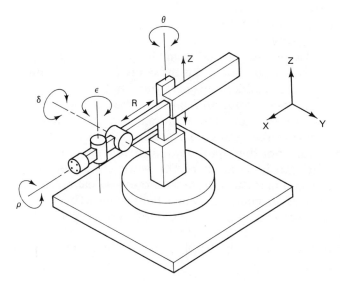

Figure 2-6 Cylindrical Geometry Robot

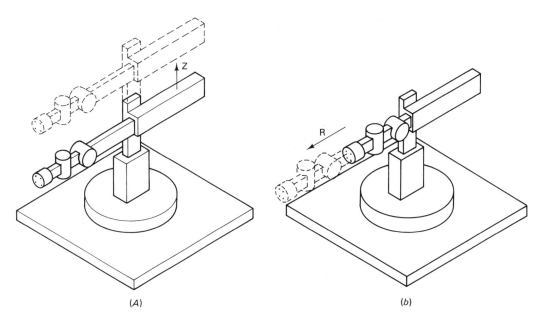

Figure 2-7a Z Axis Movement

Figure 2-7b R Axis Movement

post also requires movement along only one axis. A change in the R direction by the horizontal arm of the robot would change the gripper position without requiring motion from any other axis. This motion is described in Figure 2-7b. Movement of the gripper along the Y coordinate in Figure 2-6 requires changes in three axes in order for the gripper to maintain the same part orientation. The changes required in the three axes are described in Figure 2-8. Figure 2-8a shows the rotation about the base and the desired new position. Note that the rotation of the base causes the part to move off the Y coordinate and also causes a rotation in the part with respect to its original orientation. Figure 2-8b shows the movement that is required by the reach actuator (R) to reposition the part on the Y coordinate at point B. As shown in Figure 2-8c, rotation from the wrist in the ϵ axis is necessary to correct for part orientation. This required motion in three axes dictates a controller that can coordinate the movement of all three actuators.

The axes on cylindrical coordinate robots are driven pneumatically, hydraulically, or electrically. An illustration of the work envelope for the cylindrical machine appears in Figure 2-9; note that the base cannot rotate a full 360 degrees because of mechanical design limitations. The vertical movement H of the robot is determined by the size of the linear actuators in that direction.

Some of the advantages of the cylindrical coordinate geometry are as follows:

Sec. 2-2   Robot Arm Geometry

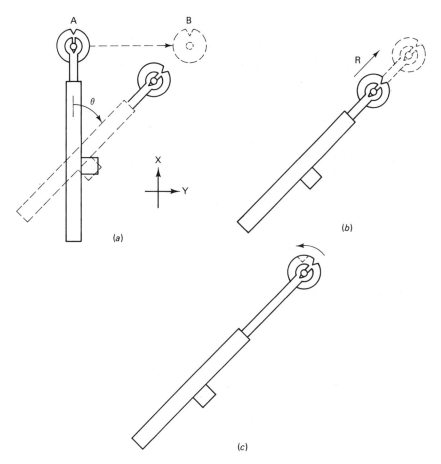

Figure 2-8 Y Coordinate Moves for Cylindrical Robot

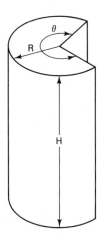

Figure 2-9 Work Envelope for the Cylindrical Robot

- Deep horizontal reach into production machines is possible.
- The vertical structure of the machine conserves floor space.
- A very rigid structure is possible for large payloads and good repeatability.

The singular disadvantage is the limited reach to left and right because of the mechanical constraints which limit the size of the horizontal actuator. This is often overcome by mounting the robot on a movable platform which can be positioned anywhere along the *Y* coordinate.

This type of arm geometry can be used for most applications but is especially desirable when deep horizontal reach is necessary or when the manufacturing layout consists of machines to be serviced by the robot in a circle with a small radius. The die casting application in Figure 2-10 is an example of this type of production layout. Examples of robots with cylindrical geometry are pictured in Figures 2-11, 2-12, and 2-13. Note the large range of maximum load capability on the machines pictured and also the movable base in Figure 2-11 that provides greater reach.

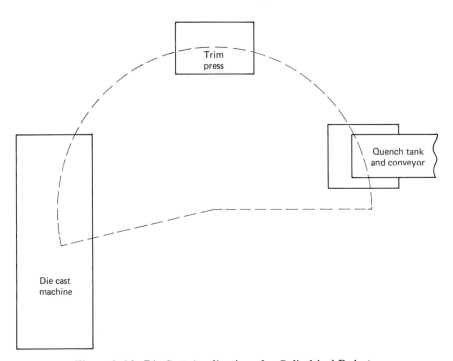

Figure 2-10 Die Cast Applications for Cylindrical Robots

Figure 2-11 Prab Robot with a Linear Base (*Courtesy of Prab Robots, Inc.*)

Figure 2-12 IBM Model 7535 Robot (*Courtesy of International Business Machines*)

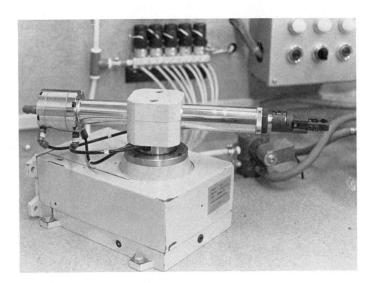

Figure 2-13 Seiko Model 700 Robot

Spherical Coordinate Robots

In the previous section you discovered that the cylindrical geometry arm required coordinated motion along more than one axis for the moves in the Y direction. The spherical geometry arm, however, will require this type of coordinated motion for movement in all three positioning directions, X, Y, and Z. Spherical arm geometry positions the wrist through two rotation axes (θ, ϕ) and one linear actuation (R). As in the previous cases, the orientation of the tool plate is achieved through three rotations in the wrist (ρ, roll; ϵ, yaw; δ, pitch). A robot employing spherical coordinate geometry is described in Figure 2-14. In theory the arm rotation (ϕ) could be 180 degrees or greater, and the waist rotation (θ) could be 360 degrees. Then if R, robot reach, went from the retracted to the fully extended position, the volume of operating space defined by the movement of ϕ, θ, and R would be two concentric half spheres. The work envelope for one of the most frequently used spherical geometry machines is illustrated in Figure 2-15. Note that the actual working volume is much less than the theoretical volume of the machine in Figure 2-14. Again, this results from mechanical design constraints. Figure 2-16 is a small spherical geometry robot designed to be mounted on a numerical control machine tool to load and unload parts. The design provides the required larger work envelope necessary for this application. Spherical geometry machines use either hydraulic or electric drives as the prime movers on the six axes, with pneumatic actuation used to open and close the gripper.

Sec. 2-2 Robot Arm Geometry

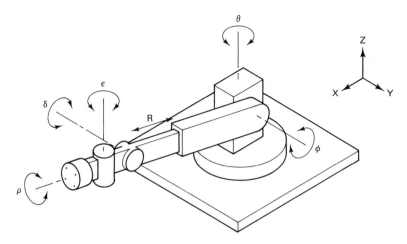

Figure 2-14 Spherical Geometry Robot

The advantages and disadvantages listed for cylindrical geometry can also be applied to spherical coordinate geometry, with the following exception: cylindrical geometry is more vertical in structure; spherical coordinates yield a low and long machine size. This is especially true for spherical machines which are designed to provide long horizontal reach.

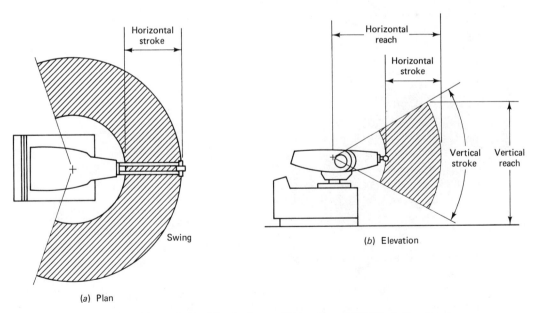

Figure 2-15 Westinghouse/Unimation 2000 Work Envelope

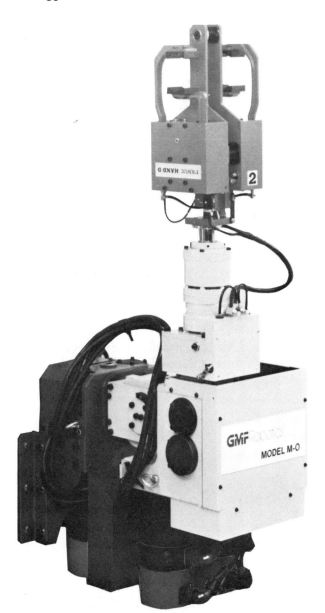

Figure 2-16 GMF Machine Tending Robot (*Courtesy of GMF Robotics*)

Jointed-Spherical Coordinate Systems

The jointed-spherical coordinate arm illustrated in Figure 2-17 is also referred to as the polar or anthropomorphic configuration. This design approximates human movements most closely, with waist, shoulder, and elbow rotation provided by the joints at θ, ϕ, and β, respectively. As in the previous arm designs the orientation of the tool plate is provided by the three rotations in the wrist. Figure 2-18 shows the work envelope of a Cincinnati Milacron T3-566 machine to be basically spherical; however, mechanical

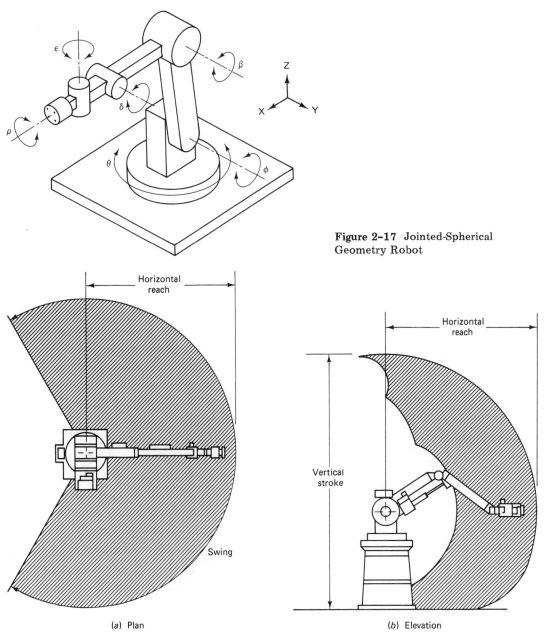

Figure 2-17 Jointed-Spherical Geometry Robot

Figure 2-18 Jointed-Spherical Work Envelope

constraints limit the actual work envelope to something less than the theoretical total half sphere. Straight-line motion along any of the three coordinates, X, Y, or Z, requires the coordinated movement of a minimum of three joints; therefore, sophisticated controllers are generally required for this type of arm geometry. The axes of most machines are driven by either electric or hydraulic power sources with feedback control systems. The human-like movements of the jointed-spherical arm create the following advantages for robotic applications:

- Although it occupies a minimum of floor space, the robot achieves deep horizontal reach.
- A good size-to-reach ratio is achieved, a result of the arm's ability to fold up when in the retracted position.
- High positioning mobility of the end-of-arm tooling allows the arm to reach into enclosures and around obstructions.

Figure 2-19 Cincinnati Milacron T3-566 Jointed-Spherical Robot (*Courtesy of Cincinnati Milacron*)

Sec. 2-3   Power Sources   39

This type robot has the drawback, however, that more sophisticated control requirements result in higher cost for the machine. The reach and size of the jointed-spherical configuration is illustrated in Figure 2-19.

2-3 POWER SOURCES

Study and work in the field of robotics offers a challenge quite unlike any other area because the design of manufacturing cells with robots draws from many technical specialties and spans numerous engineering fields. A review of power sources for robot systems supports this multidisciplinary concept. The three primary power sources used to drive manufacturing systems, namely, *hydraulics*, *pneumatics*, and *electromotive force*, are also used as prime movers in current robots.

Hydraulic Drive

A basic hydraulic power system is illustrated in Figure 2-20; study the diagram carefully. The pump and tank provide oil at high pressure for the system; four-way control valves switch the flow of high-pressure oil, and one or more actuators produce the desired motion. The two actuators commonly used are the *linear* type for straight-line motion and the *rotary* type where rotational torque is desired. In the case of the linear actuator, the high-pressure hydraulic oil is forced into one end of the hydraulic cylinder. When the chamber fills the high-pressure oil causes the piston to move; this movement forces the oil on the other side of the piston to flow out of the cylinder. This oil is returned to the pump through the four-way valve and return lines. In Figure 2-20 the solid arrows indicate the direction of oil flow and the resulting actuator movement. To reverse the actuator action, the fluid must flow in the direction of the phantom arrows. As indicated, the four-way valves, which control actuator movement, are driven by electrical signals from the robot control system. Figure 1-4 shows a Cincinnati Milacron T3-566 robot with its five rotary actuators (labeled A through E) and one linear actuator (labeled F).

Hydraulic actuators have one primary advantage: a very high power-to-size ratio affords large load capability on robots using this type of power source. That single advantage is offset by a number of disadvantages:

- Even the best hydraulic system will leak eventually.
- Hydraulic oil can become a fire hazard in arc welding applications.
- The additional equipment in the form of motor, pump, tank, and controls increases maintenance, energy, and robot costs.

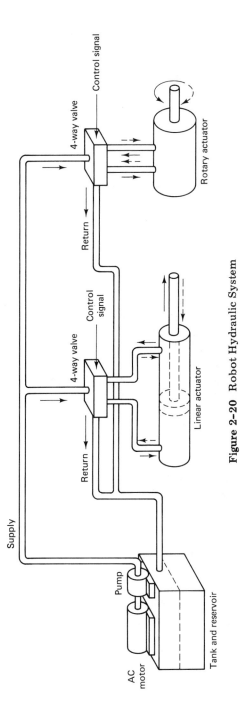

Figure 2-20 Robot Hydraulic System

- A higher noise level is associated with hydraulic systems.
- Both electrical and mechanical maintenance workers must be present for repairs in some labor situations.

Although the list of disadvantages outnumbers the advantages, a significant number of robots with hydraulic actuators have been sold because of the need for large payloads in certain industrial applications.

Pneumatic Drive

The basic system components in pneumatics are the same as those identified for the hydraulic system. The primary difference is that the power is being transferred by a gas under pressure rather than by oil. In most robotic applications, the pneumatic actuators operate with only two usable positions, retracted and fully extended, rather than using feedback to achieve proportional control. Using a feedback control system to hold a pneumatic actuator at a 50 percent extended position with the required repeatability is difficult because of the properties of the compressed gas, but the repeatability is very good if the actuator is driven against fixed stops at each extreme of travel. In some systems the fixed stops are adjustable slide blocks which stop the actuator as it extends or retracts. The gas serves to force the movable part of the actuator against the block; robot repeatability is maintained by the block setting. A robot using this fixed-stop approach is called a *pick-and-place* device or *bang-bang* machine. The positioning limitations do not limit the robot's applications, however; in fact, there are over 60,000 robots of this type presently in use in Japan.

The pneumatic power source has the following advantages:

- It is available in most manufacturing areas.
- Pneumatics is an inexpensive, well-developed technology.
- System leaks do not contaminate the work area.

The primary disadvantage has been the inability to drive the pneumatic system using feedback control to provide proportional operation and multiple stops. Work is being done in this area, however, and at least one robot is presently available with pneumatic drive and full proportional control. Figures 2-13 and 2-21 illustrate pneumatic-powered robots for small and large applications.

Electric Drive

The electric system includes a source of electrical power and an electric motor. In most applications the motors are servomotors, but stepper motors are used on some robots where the payload is small. The servomotor is

Figure 2-21 RC-4 Pneumatic Robot from PSL Robotics (*Courtesy of Platt Saco Lowell Inc.*)

primarily dc, but ac types are taking over in many Japanese models. The electric motor provides an excellent source of rotational torque either directly or through gearing, but it must rely on ballscrew drives for large, accurate linear movements. As a result many electric robots have an arm geometry which is a jointed-spherical type so that all joints can be driven by rotary drives.

The advantages of the all-electric drive include:

- No generation of hydraulic or pneumatic power is required.
- No contamination of the work space occurs.
- Low noise level is maintained while operating.

The disadvantage is the limited lifting or payload capability of the electric system compared to its hydraulic counterpart. As a result, electrically powered robots are primarily designed for parts assembly or welding and coating applications where the payload and end-of-arm tooling is 200 pounds or less. The robot in Figure 2-22 is available in three sizes depending on the payload requirements of the job; its jointed-spherical geometry uses all-electric drives.

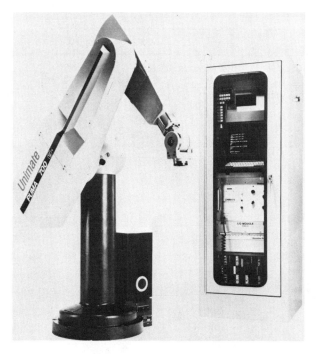

Figure 2-22 Puma Robot from Westinghouse/Unimation (*Courtesy of Westinghouse/Unimation Inc.*)

2-4 APPLICATION AREAS

It has been said that robotics is a solution looking for a problem. In the early days of robot design, any problem or application for which a company sought a solution was welcomed, and robots were adapted as much as possible to solve the task at hand. The robot industry has now come of age, and robot manufacturers are zeroing in on target applications with machines whose characteristics match the specific job. General Motors, for example, is installing paint-spraying robots which have two arms: one for spraying the paint and one for opening the door of the car. Applications can be divided into two groups: *assembly* and *nonassembly*.

Assembly

The important application area for the next ten years will be in the assembly of parts and systems by robots. Table 2-1 compares new robot applications projected for 1990 to actual implementation of robot systems in 1981. The robots that will evolve into the next generation of assemblers are at present small table top models capable of payloads of 25 pounds or less. The PUMA which is produced by Westinghouse Unimation Inc. was the first widely used robot designed specifically for assembly. The system pictured in Figure 2-23 is the first small assembly model manufactured by Cincinnati Milacron.

TABLE 2-1

	1981	1990
Spot-welding	35-45%	3-5%
Arc-welding	5-8%	15-20%
Materials handling/machine tending	25-30%	30-35%
Paint spraying	8-12%	5%
Assembly	10%	35-40%
Other	8-10%	7-10%

Figure 2-23 Cincinnati Milacron T3-726 Small Electric Robot

Nonassembly

The primary applications in the nonassembly area have been welding, spraying or coating, material handling, and machine loading and unloading. These will continue to be the major areas in the 1980s. The line drawn between assembly and nonassembly robots is certainly not a clear one, however, since robots in the PUMA series have been used in nonassembly tasks such as arc welding and material handling. Classification by applications is valid since there are tasks at which each machine excels; yet most manufacturers are unwilling to do this since it may eliminate their product from a potential sale.

2-5 CONTROL TECHNIQUES

The type of control used to position the tooling separates robots into the categories of *servo* (closed-loop) or *non-servo* (open-loop) systems. Figure 2-24 is a simplified diagram of a servo or closed-loop system with two sensors to continuously measure the angle of the joint and the rate at which it is changing. In the closed-loop control system the position of the robot arm is continuously monitored by a position sensor, and the power to the actuator is continuously altered so that the movement of the arm conforms to the desired path in both direction and velocity.

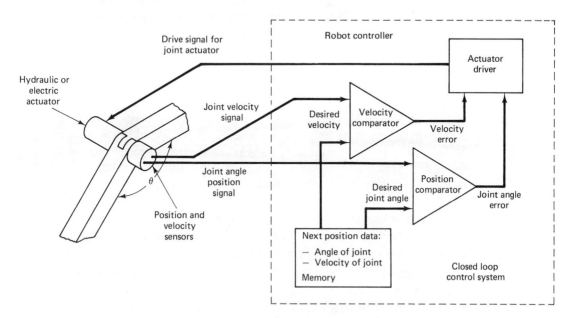

Figure 2-24 Closed Loop System

In Figure 2-24 the closed-loop control system of one joint of a six-axis machine is shown. The other five joints would have duplicate circuits. The controller gets joint angle information from a position sensor which could be a potentiometer, encoder, or resolver. Each of these position sensors can provide a continuous electrical signal which represents the angle of the joint. The current position of the joint is available from the position sensor. Thus the controller can compare it to the next desired joint angle, which is calculated from the data stored in the controller memory when the robot was programmed. This electronic comparator will produce a joint angle error signal which can be used by the actuator driver to change the joint angle so it will agree with the desired value.

In a similar manner the rate of change of the joint angle is continuously monitored by the velocity sensor, which is usually a tachometer measuring the rpm of the rotary actuator. The rate at which the joint is moving is directly proportional to the rpm of the rotary actuator driving the joint. The rate of change of the joint is compared to the desired rate of change by a separate comparator and produces a rate of change or velocity error. The error from the two comparators is analyzed by the actuator driver electronics to determine the velocity and direction the actuator should assume for the joint to reach the value required for the next position in the program.

If a machine has six degrees of freedom or six axes, then the position of the joint in each degree of freedom is individually measured, and the electrical signals are sent by wire to the controller for analysis by the servo electronics. This sampling of joint positions and corresponding corrective action to the actuator occurs many times a second so that the visual effect is a smooth movement in the robot arm from one position to another.

Closed-Loop System

The *servo* or *closed-loop* system is used in any application where path control is required, such as in welding, coating, and assembly operations. The more sophisticated controllers include a computer, display, keyboard, programming pendant, and input/output ports, in addition to the servo control system. The advantages of servo control are:

- Flexible program control permits robots to be used in a wide variety of manufacturing jobs and extends the useful lifetime of the machine.
- Robots are capable of performing more complex manufacturing tasks.
- End effectors and tooling can be less complex because of robot positioning capability.
- Robots can execute multiple programs to handle varied manufacturing tasks.

Sec. 2-5   Control Techniques

The primary disadvantages include the following points:

- Bigger capital investment for a machine of this type is required.
- Maintenance staff must be highly skilled because of the increased technology in the work cell.

The Cincinnati Milacron T3-726 robot in Figure 2-23 is an example of this type of servo feedback control system.

Open-Loop System

Robots classified as *non-servo* do not have position and rate-of-change sensors on every axis; therefore, the controller does not know the position of the tool while the robot is moving from one point to another. On every axis there is a *fixed stop* or *limit* at each extreme of travel which provides the positioning accuracy at that point. Figure 2-25 is a simplified block diagram of a *non-servo* (open-loop) system. Note that there are no signals coming from the joint or axis to tell the controller the current position or velocity of the joint. The joint will only stop moving when it reaches one of the two possible extremes of travel or when the actuator driver removes the drive signal after a predetermined amount of time. A signal from the controller causes the power source to drive the axis actuator from the present limit or stop to the opposite limit. The controller often does not verify that the

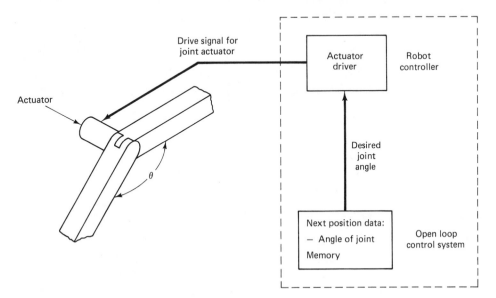

Figure 2-25 Open Loop System

new position has been reached; however, the integrity of the mechanical system and the time allowed to reach the new position assure accurate operation. It should be emphasized that very few problems occur when using open-loop control in standard applications.

A technique known as *limit sensing* is used when an application demands verification that a limit has been reached. For example, if a non-servo robot is used to unload a die casting machine, the system design would require verification that the robot gripper and the part held in the gripper are both clear of the dies before the die casting machine would be cycled and the dies would be closed. Therefore, a switch is mounted at the limit to be sensed. Next the controller drives the actuator to that limit and awaits verification from the limit switch before commanding any further axis motion. This same technique can be used to provide multiple stops or limits on an open-loop type machine. Figure 2-26 shows a rotary actuator on the base of a Prab robot with adjustable limit switches to control base rotation. As the base rotates, the actuator driver can turn off actuator power whenever a limit switch is reached and provide the non-servo system with variable stop positions.

Figure 2-26 Multiple Stops on a Prab Non-Servo Feedback System (*Courtesy of Prab Robots Inc.*)

The open-loop system is referred to in the industry as a *stop-to-stop* or *pick-and-place* type robot. A pneumatically powered modular component system is pictured in Figure 2-4. The system is assembled by the user from a wide selection of linear and rotary actuators. A second stop-to-stop robot is illustrated in Figure 2-27. The advantages of the servo type robots become

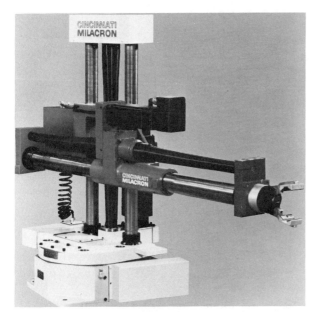

Figure 2-27 Pick-and-Place Robot (*Courtesy of Cincinnati Milacron*)

the disadvantages for the non-servo stop-to-stop type; yet despite reduced flexibility, the applications for this type robot abound in industry. The low initial cost makes this system an attractive choice for machine load-and-unload applications.

2-6 PATH CONTROL

The least ambiguous method for classifying robots is based on the type of path control which the controller provides. *Path control* is a way of defining the method that the robot controller is using to guide the tooling through the many points in the desired arm trajectory. The four types of path control from least complex to most complex are *stop-to-stop*, *continuous*, *point-to-point*, and *controlled path*. Some segments of the robotics industry identify stop-to-stop machines as point-to-point without servo feedback; however, except for this one deviation the industry is consistent.

Before describing the four types of path control, a clear understanding of how the controller is storing the program points is important. Figure 2-28 shows a simplified robot with four degrees of freedom programmed to grasp a part. The robot is shown in the three positions through which it must pass in order to complete the pickup of the part. Number one is the starting point; point two positions the gripper over the part; and the third point puts the gripper in a position where the parallel jaws can close to secure the part. For this operation the three points represent the minimum number of pro-

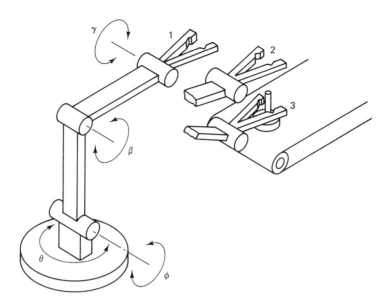

Figure 2-28 Point-to-Point Path Control

gram steps which would have to be used. The robot operating system would record in the controller memory the necessary information about the path to permit the robot to return to these three points as the program is executed. The minimum information necessary to permit the robot to return to any point in its work envelope is the position of each degree of freedom or of each axis when the robot is at the point. In the case of joints that move through rotation, the angle is the critical value, and in the case of linear movement the distance extended would have to be preserved. The actual data stored in memory for each programmed point vary, depending on the type of path control present. Continuous-path type machines store the actual joint angles for each programmed point. For point-to-point servo machines the system stores the Cartesian space coordinates of the programmed points. For example, the X, Y, Z, δ, ϵ, ρ values would be saved. These values are used to calculate joint angles or actuator extensions during program execution. Figure 2-29 shows a portion of the memory of a continuous path system with joint angles for three points recorded. In actual practice the information in memory would be recorded as binary (1s and 0s) numbers; the decimal values are used for clarity. Therefore, regardless of the path control in use by a robot system the problem is reduced to recording the critical information about every point in the work envelope that the robot must pass through in the process of executing a series of moves. This basic process applies to each type of path control discussed in this chapter.

Sec. 2-6   Path Control   51

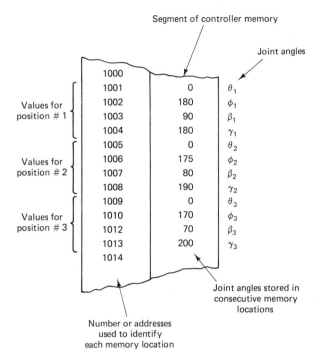

Figure 2-29 A Segment of a Robot Controller Memory

Stop-to-stop

In *stop-to-stop path control* the robot system is operating open-loop, which means that the position and velocity of the axis is not known to the controller. An example will demonstrate this. The robot in Figure 2-30 must go from point A to point B as part of a programmed move. Since the motion of the actuators is not sampled with a feedback system, the actual position of the axis is not known until the actuator is driven to the desired limit, which is determined by the mechanical stops in the actuator cylinders. As a result, the only information stored in memory is a sequential list of *on/off* commands for each actuator driver. In this case the memory would have actuator one *on* followed by actuator two *on*.

Point-to-point

The primary programming device in *point-to-point controllers* is the hand-held teach pendant. The controls on the pendant for the IBM 7535 robot controller are shown in Figure 2-31. Teach pendants usually have two momentary pushbutton switches for each degree-of-freedom on the robot. The programmer can move each axis independently in either direction with these

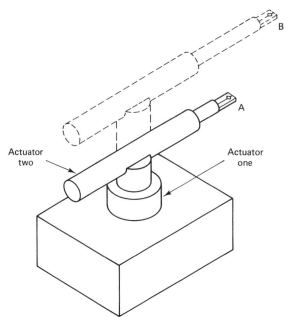

Figure 2-30 Stop-to-Stop Path Control

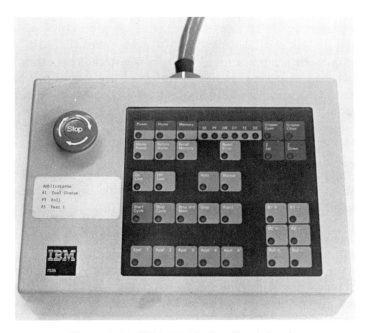

Figure 2-31 IBM 7535 Robot Teach Pendant

controls. In addition, the pendant provides a programming button which commands the controller to record in memory the current position of every joint or axis on the robot. A button is also provided for emergency stop and single stepping through a program. These functions are common among the teach pendants of most manufacturers; often, additional control switches are provided for special functions of a machine, such as speed control during programming.

The tooling or gripper is moved into position with the teach pendant; then the robot controller is commanded to record the position of the tooling in memory. This information represents one point in a programmed move. The same procedure is followed at the next desired point in the work cell when the positions of all the joints are feedback to the controller to calculate the Cartesian space coordinates for the tool position. Finally, these X, Y, Z, δ, ϵ, ρ values are recorded by the system when the record button is depressed. This process is repeated until all the desired positions of the tool have been stored in the controller memory in the form of Cartesian space coordinates. It is important to note that the path which the robot takes as the programmer moves the tool from one desired position to another will have no effect on the final program path when the program is run. Figure 2-32 illustrates this concept with the positions in space which would be required for a point-to-point path controller to move a part from one conveyor to another. The following four points would be required to pick up the part at point 1; raise it to point 2; transfer it to point 3; and lower it to point 4. The solid lines represent the desired path of movement when the program is executed by the controller. The dotted line represents the path that the part might have taken when the programmer was teaching the robot the desired operation. *The points marked one through four are the only points that were recorded in the controller memory by the programmer,* so the deviation from the desired path as the programmer moved the tool from point one to point two did not affect the final operation.

Point-to-point controllers can store thousands of program points for a complex application and can also store several different programs; thus a robot can adapt its movement to different manufactured parts or conditions in the work cell. For example, in an inspection operation the controller could drive the gripper through a programmed routine that moves parts through a laser gauge and onto an output conveyor. As soon as a part fails inspection, however, the process must be altered to drop the part into a scrap bin. This is accomplished by activating a second stored program which has the set of points for the path to the scrap bin. This process is called *branching.* The Unimate 2000 is an example of the point-to-point type robot system with a storage capability of 2048 programmed points.

The point-to-point type system provides feedback control of each axis over the entire path, with the controller driving each axis from its starting

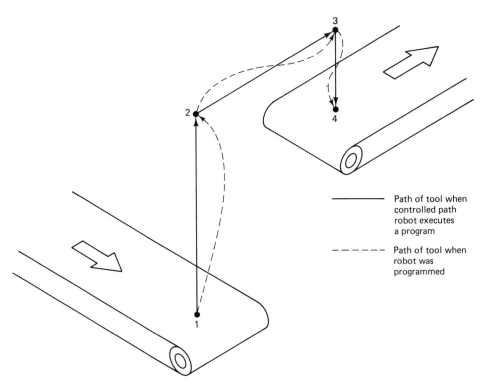

Figure 2-32 Point-to-Point Programming Example

angle or position to the angle required for the next programmed point along the path. Figure 2-33 shows a symbolic spherical robot moving from point 1 to point 2, where the first and second points are the same distance from the robot, but point 2 is higher. The controller must change four degrees of freedom in this example. The base must rotate through angle θ; the shoulder

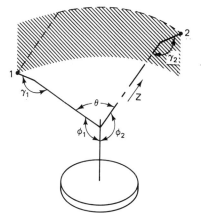

Figure 2-33 Example of Point-to-Point Non-straight Line Motion

must go from angle $\phi 1$ to angle $\phi 2$; the arm must extend along Z to reach the higher point; and the wrist must pitch from angle $\gamma 1$ to $\gamma 2$. The controller will change each axis at its maximum rate; therefore, the shoulder and pitch actuators, with the least change, will complete the move well in advance of the base actuator, which requires the greatest travel distance. This causes the end effector to assume the nonstraight-line path indicated by the dotted line in Figure 2-33. This lack of path control presents no problems with robots used in applications such as material handling or machine tending where the straight-line path is not required; however, it could cause programming difficulty in arc welding situations where the robot is required to follow a seam between two metal parts. In order to achieve straight-line motion with a point-to-point controller of this type a large number of points must be programmed very close together to keep the changes in every axis small and about equal.

The advantage of point-to-point type control is that relatively large and complex programs can be obtained with a system that is moderate in cost yet has proven reliability. Although there are many applications where path control is not a requirement, the lack of straight-line control does limit the number of applications of a point-to-point system and would thus be considered its primary disadvantage.

There is a very large selection of servo-controlled point-to-point robots available with either electric or hydraulic drives and with any desired arm geometry; however, the jointed-spherical type is the most popular. The lifting capacity varies from several kilograms to over 900 kilograms with repeatabilities as precise as plus or minus 0.05 millimeters.

Controlled Path

The *controlled path* robot is a point-to-point system with added capability to provide control of the end effector or tool-center-point as it moves from one program point to another. The system is programmed in the same manner as the standard point-to-point machines with each point in the path recorded using the teach pendant. The difference occurs when the program is executed, with the primary distinction being the *straight-line motion* between the programmed points. The axis actuators are driven in a proportional manner, with the axis requiring the most change driven faster than the axis requiring the least change. One result is that a straight path is followed between programmed points without any additional programmer dexterity. In addition, the *velocity* between points can be individually specified in the program, along with special tool moves such as an arc welding weave pattern. All this capability is a result of increased controller intelligence and has an added benefit of enhanced programming features such as program editing, trouble diagnostics, larger memory capability, and added end effector control.

Continuous Path

The primary difference between point-to-point controlled path just described and *continuous path* control is the number of programmed points which are saved in controller memory and the method used to save them. The programmed path in Figure 2-34 illustrates this difference. The point-to-point type of robot would require just four programmed points stored in memory to record this arm motion. However, the continuous path machine would store hundreds of points for the same arm motion.

The reason for the large difference in the number of points stored is the method used by each type system to record a programmed path. In point-to-point programming the programmer moves the robot to the desired location and presses the program button. A single point in the programmed path is saved. After a sequence of these the program is complete. The point-to-point controller then drives the robot from one point to the next. By

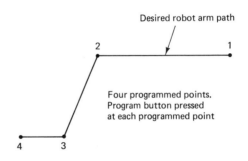

(a) Point-to-point type control

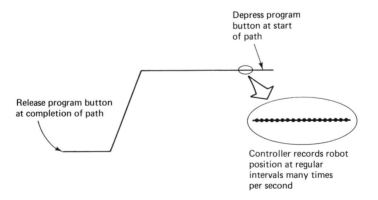

(b) Continuous path control

Figure 2-34 Comparison of Point-to-Point and Continuous Path Type Control

contrast, in continuous-path programming the programmer teaches the desired path using a teach stand or by physically moving the robot arm. The program button is depressed at the start of the move and not released until the desired path is completed. While the program button is depressed the continuous-path controller is recording program points in memory at the rate of six or more per second.

The continuous-path programming process records every move that the programmer makes as the arm or teach stand is moved. The point-to-point process, however, records only the location of the arm when the program button is pressed. Any motion in the arm when it is moved from one point to the next is not recorded. Continuous-path controllers are very useful in applications like paint spraying where the robot's motion must duplicate the skill of the operator.

2-7 CONTROLLER INTELLIGENCE

Our last classification technique separates machines into three groups by means of the relative intelligence level of the robot controller; the three groups are *high* technology, *medium* technology, and *low* technology. As the industrial robot industry developed, there were a large number of mechanically simple machines which performed relatively easy tasks such as machine tending. Robots of this type, used frequently by the Japanese, did not require complex mechanical design or a sophisticated controller so they were grouped into the *low*-technology category. Some of the older robots, such as the Versatran and Unimate series, were more versatile than the simple pick-and-place machines so they were grouped together in a category called *medium* technology. As the more sophisticated systems started reaching the market, a new category labeled *high technology* was formed.

Today, the line separating the three groups is less defined because the low cost of computing power makes it possible for every system to have a sophisticated controller. It is easy to identify some low-technology systems which are non-servo and are applied to single manufacturing tasks. At the same time, high-technology robots can also be found which have good repeatability, proportional control, and the capability to control the operation of the entire work cell. As a result, classification of systems as either high technology or low technology is easier to defend.

2-8 SUMMARY

The objective of this chapter has been to introduce the general concept of robot classification and to provide an overview of all types of robot systems. Industrial robots can be grouped by several techniques which include *arm* geometry, *power source* for the joint actuators, intended *applications*, the presence or absence of *feedback*, the type of *path control*, and the *controller intelligence*.

The arm geometries currently available include *rectilinear* coordinate, *cylindrical* coordinate, *spherical* coordinate, and *jointed-spherical* coordinate. These mechanical configurations can use *pneumatic*, *hydraulic*, or *electric* drives as a source of power for the joints or axes that are moved. Application is not a good classification technique, but generally the areas of painting/coating, arc welding, assembly, and machine tending have received special attention from the robot industry. A much better classification scheme designates a robot system as either *servo* or *non-servo* controlled, which means that the system either knows the current position of every axis through the use of feedback signals or it does not sample the current position values and has no feedback. Another good classification technique uses path control to distinguish between current robots. The path control types include *point-to-point*, *controlled path point-to-point*, *continuous*, and *stop-to-stop*, with the last type utilizing open-loop control in most situations. Another possible classification method rates the intelligence level of the controller. *Low-*, *medium-*, and *high*-technology robots were defined by the industry in early machines, but with the current type of controllers the middle category is difficult to isolate.

QUESTIONS

1. On the basis of work experience or from industrial trade magazines, identify three applications for each type of arm geometry. Select applications which would best suit each type of arm geometry.
2. Compare the three types of drive power sources by preparing a table which illustrates the strengths and weaknesses of each.
3. Make a list of the most important machine characteristics which should be incorporated in an assembly robot system.
4. What is the primary difference between a servo and non-servo robot system?
5. What types of manufacturing applications would be best served by a non-servo system?
6. What characteristics of manufacturing applications require that the robot used have a closed-loop system?
7. How does the controlled path point-to-point system differ from the standard point-to-point system in the operation mode? How do they differ in the program mode?
8. Why does the continuous-path system have a limitation on the length and number of programs which can be stored?
9. Why is the continuous-path system ideal for applications such as paint spraying and coating?

Chap. 2   Questions   59

10. List three typical applications for each of the four types of path control. Try to match job tasks with the strengths of each type of path control.
11. The Cincinnati Milacron T3-566 can be described as a six-axis jointed-spherical arm with hydraulic drive and servo driven controlled path point-to-point operation. Use industrial trade and robot magazines to write a similar description of ten other robots, listing the number of axes, arm geometry, type of feedback, and the path control.

Handwritten notes:

Non Controlled — machine tending, die casting, parts manipulation
Controlled path — inspection, assembly, tool manipulation
Continuous path — coating, spraying, welding

RECTANGULAR — palletizing, parts handling, assembling small parts
CYLINDRICAL — small parts handling, machine loading & unloading
SPHERICAL — palletizing, machine loading & unloading
JOINTED SPHERICAL — welding, spraying, assembly

HYDRAULIC
 Advantages: High power, Large Load
 Dis-Advantage: Leakage & Fire Hazard, Extra equipment

PNEUMATIC
 A — Power is there usually, inexpensive, clean
 D — No proportional control

ELECTRIC
 A — clean, quiet
 D — limited payload

3

END-OF-ARM TOOLING

3-1 GENERAL CHARACTERISTICS

Studies indicate that the process of joining two mechanical parts in an assembly operation utilizes an operator's sense of touch to a much greater degree than the sense of vision. This dominance of the tactile underscores the importance of the human hands in all phases of manufacturing and assembly of production goods. It also emphasizes the demand placed upon the *gripper* or *end-of-arm tooling* of an industrial robot if it is to perform many of the production duties of its human counterpart. Duplication of the human hand with its ability to grasp, sense, and manipulate objects remains one of the most difficult tasks facing the designer of end-of-arm tooling. Both Japanese and American designers are working on a robot hand with three fingers which can grasp irregularly shaped objects, but application of this type of tooling is several years away. This chapter investigates the types of robot tooling currently in use and covers the terminology involved in end-of-arm or *end effector* tooling.

In general, an end effector is the tooling or gripper which is mounted on the robot tool plate. The function of the gripper is to hold the part as the robot presents it to the tool for work to be done, or to hold the tool as the robot moves it to work on the part. For example, a part held in a gripper could be positioned under a numerical control (NC) drill, thereby producing a hole; or the end-of-arm tooling could be the drilling mechanism itself, with the robot producing the holes in the same manner as a human operator using a hand drill.

The end-of-arm tooling used in a robot work cell should exemplify all five of the following characteristics:

1. The tooling must be capable of gripping, lifting, and releasing the part or family of parts required by the manufacturing process.
2. The tooling must sense the presence of a part in the gripper, using sensors located either on the tooling or at a fixed position in the work cell.
3. Tooling weight must be kept to a minimum since it is added to part weight to determine maximum payload.
4. Containment of the part in the gripper must be assured under conditions of maximum velocity at the tool plate and loss of gripper power.
5. The simplest gripper which meets the first four criteria should be the one implemented.

3-2 CONSERVATION OF JOB IQ

The Law of Conservation of Energy exists in all physical systems. This law states that, within a system, energy in all of its many forms remains constant. If energy in one form increases, then energy in another form decreases, so that the total energy content of the system remains unchanged. This concept can be used to gain insight into robot system selection and end effector design or selection. When applied to robot work cell applications it is called *Conservation of Job IQ*.

Each sequence which must be performed by the robot to accomplish a specific manufacturing task has a required level of intelligence, called Job IQ. The sequence with the highest IQ or required intelligence determines the necessary intelligence level that the entire manufacturing system must have in order to make the manufactured part. For example, picking up a part from a prepositioned parts feeder requires less machine intelligence than selecting a part from a pallet of randomly located parts. Therefore, the system that loads the prepositioned parts will be a less intelligent system than the one which loads from the pallet. The greater the IQ of the manufacturing task, the greater the cost to implement the automated system. In the previous example, a low-technology robot and simple end-of-arm tooling, which together make up a rather low intelligence system, could be used to load the accurately prepositioned parts. In contrast, the randomly located parts would require a high-technology robot system with simple gripper, or a low-technology robot with an X-Y table to position the random parts for pick-and-place robot loading. In addition, the low-technology system with the X-Y positioning table may require more complex tooling to locate slightly off-center parts. The higher IQ load sequence requires a system with more intelligence, either in the form of a high-technology robot or in the form of a low-technology robot with positioning devices and complex end-of-arm tooling. In the final analysis, the sequence with the higher IQ requires hardware with a higher price tag, no matter how the problem is solved.

How does all of this apply to tooling design and selection? In setting up any manufacturing system some criteria must be used in the selection of equipment. When robots are included in the automation, an analysis of the IQs of each sequence will help establish the system intelligence level which is demanded by the manufacturing process. Every sequence in the manufacturing process should be examined, and the relative IQ of each sequence established. The system intelligence, including the end-of-arm tooling, must then equal the IQ of the most demanding sequence in the process. In this way, the robot automation intelligence level will always match the needs of the production requirement.

There is another advantage to analyzing the requirements of each sequence in the manufacturing process. Often only one sequence with a high IQ increases the manufacturing system intelligence and requirements, and consequently increases the system cost. As a result of such an analysis, the sequence with a high IQ can be identified and possibly modified to permit a much more cost-effective implementation.

3-3 CLASSIFICATION

The end-of-arm tooling used on current robots can be classified in the following three ways: (1) according to the method used to hold the part in the gripper; (2) by the special-purpose tools incorporated in the final gripper design; or (3) by the multiple-function capability of the gripper. The first category of gripping mechanisms includes standard pressure grippers, tooling utilizing vacuum for holding or lifting, and magnetic devices. The second classification of tooling includes drills, welding guns and torches, paint sprayers, and grinders. The third type of gripper tooling includes special-purpose grippers and compliance devices currently in use.

Standard Grippers

Standard grippers can have two different closing motions, *angular* or *parallel*, and can have *pneumatic, hydraulic, electric,* or *spring* power for closing and opening. The action of the angular and parallel devices is illustrated in Figure 3-1; two standard off-the-shelf grippers are pictured in Figures 3-2 and 3-3. In many applications a standard angular or parallel gripper is purchased; next, special adaptors or plates are attached to satisfy the parts-handling requirement of the robot project. In-house design of the end-of-arm tooling should only be initiated when standard purchased hardware cannot be adapted for the requirement. The cost of off-the-shelf grippers will be between four and eight percent of the basic robot cost. However, if tooling is completely designed and built in-house, the cost of the gripper can be as much as 20 percent of the total system cost.

The gripper must be closed and opened by program commands as the robot moves through the production operation. The robot controller supplies the signals which result in the gripper's action. On one type of gripper,

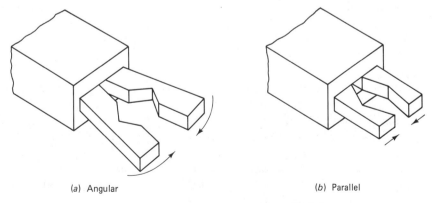

(a) Angular (b) Parallel

Figure 3-1 Standard Angular and Parallel Grippers

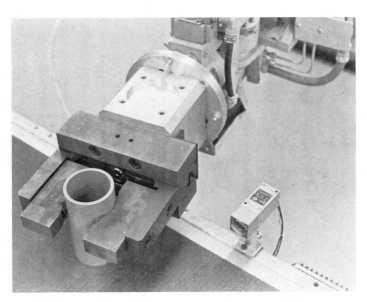

Figure 3-2 Fibro Manta Gripper Air Open and Air Close

Figure 3-3 Angular Gripper Air Close Spring Open

the tooling is *power opened* and *spring closed*, which means that the gripper is normally in the closed position, with the closing force supplied by an internal spring; in this case the gripper will only be open when power is applied to the gripper. The opposite configuration is *power closed* and *spring opened*; the final type available is *power opened* and *power closed*. Each of the types has advantages and disadvantages based on the specific application. The power closed and spring opened type, for example, only requires power to close the gripper when a part is lifted and moved, which is an advantage of simpler control and less power consumption. However, a part may be dropped or thrown if gripper power is lost while the robot is in the process of moving the part. Figure 3-4 shows a brass bearing being assembled in two different applications; the tooling in Figure 3-4a must grip on the exterior of the part, and a power closed and spring opened gripper is required. In Figure 3-4b an interior grip is necessary; therefore, a power opened and spring closed type is used.

Pneumatic power is used in most applications to open and close the jaws of the gripper when a tool command placed in the robot program is

Sec. 3-3   Classification 67

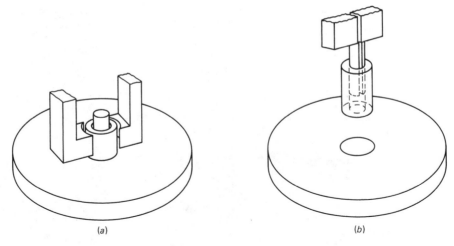

Figure 3-4 Interior and Exterior Pick-up of Round Bearing

executed. A hydraulically powered gripper is used when higher clamping force is required, as in large, heavy loads. This type is usually used on hydraulically powered robots since a hydraulic power source is already at hand. Electrically powered jaws are also used in limited applications with a solenoid or dc servomotor providing the opening and closing action. Servo drives will be used more frequently when grippers are developed which can vary the pressure applied to the object as it is grasped. This would permit the same gripper to lift either a steel ball or an egg without dropping the ball or breaking the egg. However, tactile sensing must be developed beyond its present state before this type of gripper will be practical for the factory. Figure 3-5 illustrates a large internal gripper, and Figure 3-6 shows a large parallel device.

Figure 3-5 Angular Gripper for Internal Pick-up from GCA Corporation (*Courtesy of GCA Corporation*)

Figure 3-6 Parallel Gripper on Cincinnati Milacron T3-566 Robot (*Courtesy of Cincinnati Milacron*)

Vacuum Grippers

Vacuum is used as the gripping force in many tooling applications. The part is lifted by *vacuum cups*, by a *vacuum surface*, or by a *vacuum sucker gun* incorporated into the end-of-arm tooling. The lifting power is a function of the degree of vacuum achieved and the size of the area on the part where the vacuum is applied. Since a perfect vacuum is difficult to achieve economically, the lifting capability in this application is best controlled by varying the amount of surface area of the part exposed to the vacuum.

The most frequently used vacuum gripper uses suction or vacuum cups to hold the desired part. The gripper can have a single vacuum cup, as illustrated in Figure 3-7, with a gripper picking up sheet metal plates from a stack. A proximity sensor is included to tell the robot when it has reached the top of the stack so that the robot can stop, and the vacuum on the suction cup can be activated. This system is capable of moving a variety of plate sizes in stacks of varying heights. Figure 3-8 shows a gripper with a multiple pattern of pickup cups that could lift large sheet metal plates, plastic sheets, plywood panels, or even large cardboard boxes. If the individual vacuum cups are controlled separately, the gripper can lift various size sheets by activating only the cups required for the sheet being lifted. It is this type of system flexibility inherent in a robot cell that makes it ideal for the automated factory.

The lifting capacity of the vacuum type of gripper is directly related to the pressure of the air surrounding the vacuum cups on the gripper. Figure 3-9 provides additional background on the principles involved in this

Sec. 3-3 Classification

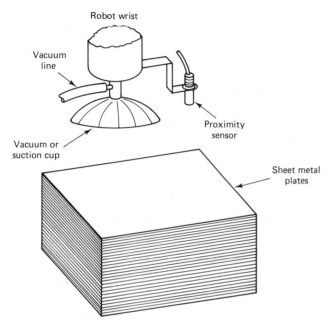

Figure 3-7 Vacuum Cup System to Unstack Sheet Metal Plates

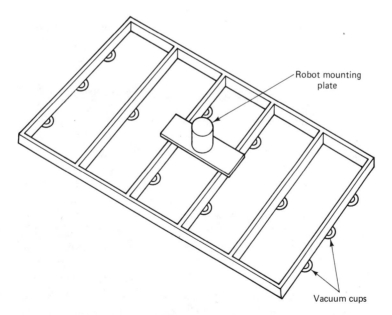

Figure 3-8 Multiple Vacuum Cup System to Handle Large Sheets of Material

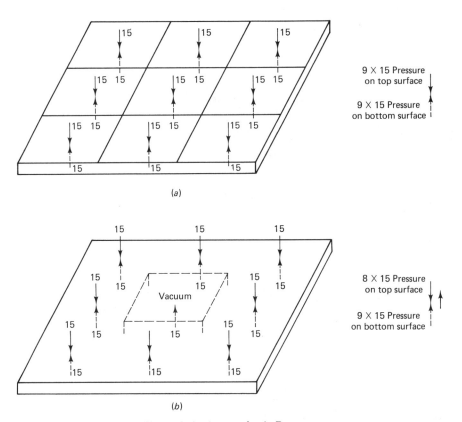

Figure 3-9 Atmospheric Pressure

lifting technique. The *atmospheric pressure* at sea level on the surface of an object is 14.7 pounds per square inch, but in this example use 15 pounds per square inch. Assuming that the thin sheet metal plate in Figure 3-9a is under normal atmospheric pressure, the force on each side of the metal is 9 square inches times 15 pounds of pressure, or 135 pounds. Now if the vacuum cup in Figure 3-9b occupies one square inch of surface on the top side, then the effective force on the top surface is only 8 times 15, or 120 pounds. The force on the bottom surface is 15 pounds greater, and the sheet metal will be held against the vacuum cup by the difference in the forces. For example, a 3-inch diameter suction cup has 7.07 square inches of surface area and could, under perfect conditions, lift approximately 15 pounds for every square inch, or 106 pounds. However, with less than a perfect vacuum and seal between cup and surface, the actual lifting capability is reduced. The surface of the material being lifted has the greatest effect on the total weight that can be lifted. The smoother the finish on the surface, the closer the actual lifting weight is to the maximum. Figure 3-10 illustrates one type of vacuum cup available.

Sec. 3-3   Classification

Figure 3-10 Vacuum Gripper
(*Courtesy of GCA Corporation*)

Vacuum Surfaces

Vacuum surfaces are just an extension of the vacuum cup principle. In some material handling applications the product to be lifted is not ridged enough for suction cups to be effective. For example, robots are used to make composite material by building up multiple layers of graphite fiber cloth and resin. To lift the cloth into place a vacuum surface such as the one illustrated in Figure 3-11 is used. Note the flat surface which comprises one side of a

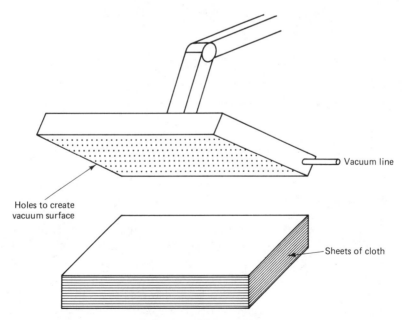

Figure 3-11 Vacuum Surface

vacuum chamber. Each hole in the vacuum surface provides a small lifting force so that the flexible cloth would be held from many points against the vacuum surface as it is moved into place. Any flexible material not too porous can be lifted effectively in this manner.

Vacuum Suckers

The use of *sucker guns* in the nylon industry is well established. The tool consists of a wand which is capable of sucking up a nylon end or a thread line as it is produced, thus permitting the operator to thread the machine. This type of end-of-arm tooling could be very useful for applications where robots must handle material in the form of thread, line, or fine wire.

Magnetic Grippers

Parts which contain ferromagnetic material can be lifted with an electro-magnet mounted on the robot tool plate. Figure 3-12 is a picture of a *magnetic gripper* available from GCA Corporation.

Figure 3-12 Magnetic Gripper (*Courtesy of GCA Corporation*)

Air Pressure Grippers

Fingers, mandrel grippers, and *pin grippers* form a group which uses air pressure to grip parts. The *fingers*, also called pneumatic fingers, have a hollow rubber-like body with a smooth surface on one side and a ribbed surface on the opposite side. With pressure applied to the inside of the hollow body, the finger deflects in the direction of the smooth side. Figure 3-13 illustrates this process, and Figure 3-14 shows a part being lifted by a robot gripper with two fingers.

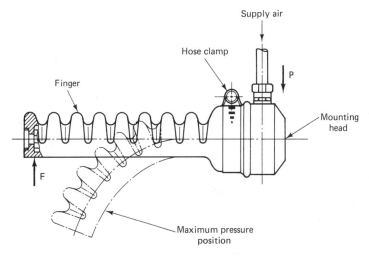

Figure 3-13 Pneumatic Finger

Figure 3-14 Finger Gripper (*Courtesy of GCA Corporation*)

Mandrel grippers are inside grippers with an airtight flexible diaphragm mounted to a mandrel. Figure 3-15 shows a section view of a circular mandrel gripper inside a part. When air under low pressure (25 to 35 psi) is forced into the port, the diaphragm expands and traps the part. Figure 3-16 shows a mandrel gripper on a robot lifting a part.

Pin grippers, Figure 3-17, are similar to mandrel grippers except that the part is gripped from the outside. The gripper fits down over round pins, and then the diaphragm is expanded inside the ridged housing and the part is trapped.

The primary advantage offered by air pressure grippers is the gentle variable force applied to the part being held.

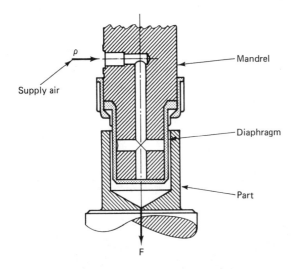

Figure 3-15 Mandrel Gripper

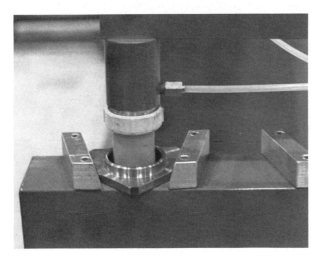

Figure 3-16 Mandrel Gripper
(*Courtesy of GCA Corporation*)

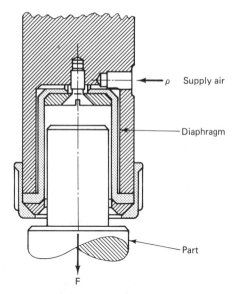

Figure 3-17 Pin Gripper

Figure 3-18 Cincinnati Milacron Robot Loading Refrigeration Liners (*Courtesy of Cincinnati Milacron*)

75

Special-Purpose Grippers

As the application of robots becomes more widespread, manufacturers will develop off-the-shelf grippers to fill special applications. GMF has begun this movement with the development of a set of grippers designed to handle circular parts for loading turning centers and lathes.

The largest classification of grippers now in use falls into the category of special-purpose tooling. In addition to those available off the shelf, many more are fabricated by the robot end user. Often the user starts with a basic pneumatic or electric mechanism available in the industry and then modifies the gripper to do the special job. In almost every case special gripper fingers must be fabricated to hold parts for specific jobs. Figures 3-18 and 3-19 illustrate some special-purpose grippers.

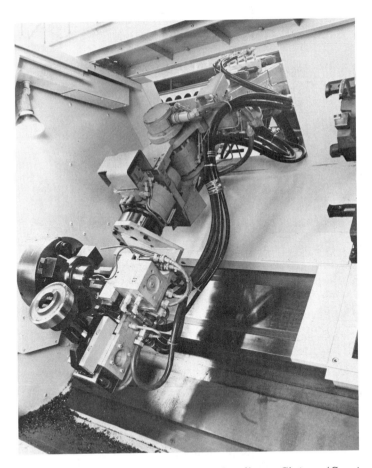

Figure 3-19 Cincinnati Milacron Robot Loading a Cinturn (*Courtesy of Cincinnati Milacron*)

Again robot flexibility is responsible for the large variety of gripper designs. It is often a simple matter to identify a secondary task which the gripper can perform during the production cycle in addition to the primary parts-handling or assembly responsibility. Examples of secondary tasks which can be incorporated into a standard gripper include hooks for lifting or turning parts over, air guns for cleaning, and oil dispensers for lubrication of parts and dies.

3-4 SPECIAL-PURPOSE TOOLS

Grippers are also designed to hold power tools in the same way that human hands do. Those most frequently used include drills, welding guns, glue and sealer dispensers, spray guns, grinders, and sand blasters. Robots are often used for jobs possessing the three D's: *dirty, dangerous,* and *dull.* A human operator using any of these tools must wear equipment to protect one or more of his/her senses, such as seeing, hearing, or breathing. As a result, these jobs are ideally suited for robots. The following studies describe typical applications using special-purpose tools:

The ability of robots to perform *drilling* tasks is a direct result of their positioning accuracy. In addition, their lifting capability provides a choice of either holding the drill in the gripper with the workpiece mounted in a fixture or picking up the part in the gripper and taking it to the drill, which is mounted in a fixed position. Current literature reports an operating work cell at General Dynamics in which a robot with a drill mounted in the end effector drills the hole pattern in aircraft cockpit canopies. The robot is capable of maintaining the drill perpendicular to the surface along the edge of the curved canopy.

Welding guns are one of the most frequently used special-purpose tools. Initially, spot or resistance welding was popular, especially in the automotive industry. Currently arc-inert gas welding systems are being used in increasing numbers. Figure 3-20 shows a Cincinnati Milacron T3 robot welding a base assembly for a computer mainframe. Although the job requires 44 seams two inches long from a variety of angles, it takes under 12 minutes to complete. In the manual mode, the assembly took 45 minutes, which is a per part time saving of over 33 minutes. In addition, seam-tracking systems are available which permit the robot to adapt its motion to variations in the seam which is being welded. Some tracking techniques use vision systems while others use welding parameters such as arc current or voltage to maintain proper tracking. Complete robot welding systems are available from the major robot manufacturers and from welding equipment companies in the United States, Europe, and Japan.

The application of *glues* and *sealers* is time-consuming and tedious when performed by hand, but an industrial robot does a perfect job repeatedly and at a high speed. A robot offers the advantages of accurate bead placement and high application rate. The consistency in rate of application provides an additional saving in material that enhances the use of robots.

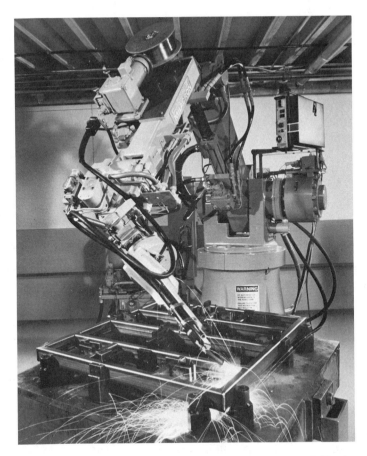

Figure 3-20 Robot Arc Welding Application (*Courtesy of Cincinnati Milacron*)

When there are enough applications of a specific type, the proper hardware will surely follow. This was never more true than in *coating* applications. The three D's aptly describe the workplace for the operator of a paint spray gun and sand blaster. Here, too, since existing robots did not satisfy all the needs of this application, several manufacturers introduced a robot specifically for coating jobs. Figure 3-21 shows a Trallfa coating robot distributed by the DeVilbiss Company in the United States.

Among dangerous workplaces, the foundry ranks very high, since the required *grinding* of the castings is a dirty and dangerous task. In this environment robots perform grinding and material handling jobs. The ASEA robot shown in Figure 3-22 is rough grinding raw castings using the grinder-sensing system incorporated in the robot. The grinder-sensing system allows the robot to sense the irregularities along the surface of a casting as the grinding is being performed.

Sec. 3-4　Special-Purpose Tools

Figure 3-21 Coating Robot (*Courtesy of DeVilbiss Company*)

Figure 3-22 Robot Grinding Application (*Courtesy of ASEA Inc.*)

3-5 ASSEMBLY FIXTURES

Assembly applications for robots will grow at a rate faster than all other current or future application areas simply because assembly requirements occur in every segment of manufacturing. As gripper dexterity and part manipulation capability increase with developing technology, and as microprocessor systems are designed to control more complex grippers, the number of assembly applications will leap at an exponential rate. Figure 3-23 shows a T3-726 robot using a pneumatic screwdriver.

Grippers designed to be used in assembly operations have special needs which must be satisfied. Parts must be moved into place for the assembly operation, fasteners must be put into place, and assembly tools must be applied. In all of these operations the *compliance* between the parts must be considered. Compliance deals with the relationship between mating parts in an assembly operation. For example, if a pin must be fitted into a hole, alignment between the hole and the pin must be achieved. Since it is impossible to guarantee perfect alignment between the robot gripper which holds

Figure 3-23 Assembly Operation with a Pneumatic Screwdriver (*Courtesy of Cincinnati Milacron*)

Sec. 3-5   Assembly Fixtures

the pin and the fixture holding the part, compliance techniques must solve the slight misalignment problems between mating parts during assembly.

Compliance

Compliance means initiated or allowed part movement for the purpose of alignment between mating parts. The two basic methods of compliance are *active* and *passive*. For each of the two methods three possible compliance conditions may be present. The initiated or allowed movement will be *lateral*, *axial*, or *rotational* about a center of rotation. Figure 3-24 shows the three basic types of movements and the misalignment which must be eliminated by each. In Figure 3-24a lateral part movement is required for mating, and in Figure 3-24b a rotational correction is necessary for proper part insertion. The axial compliance illustrated in Figure 3-24c is not neces-

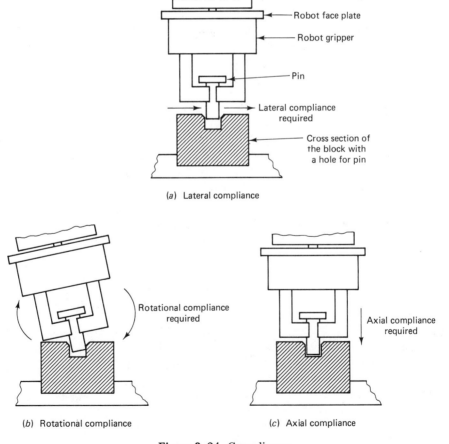

Figure 3-24 Compliance

sary for proper alignment but is necessary if the pin must be flush with the bottom of the hole. Thus the axial compliance illustrated here would operate in compression but not in tension. In compression the robot could push the part slightly past the point where the pin is seated. In tension the system could be rigid, since compliance or movement in axial tension or pulling is not required in this application. In the example, each compliance condition was considered separately, but in operation they may all be present at one time and the compliance technique employed must continue to function. Without compliance the entire system is rigid, so that additional force by the robot will result in damage to the two mating parts.

In *active compliance systems* the forces caused by part misalignment are measured by sensors and the degree of misalignment in every direction is transmitted back to the robot controller. The gripper is then moved by the controller to correct for the misalignment. Strain gauges on the gripper are used to measure the alignment forces so that the robot can move the gripper to eliminate any misalignment in the assembly process. Figure 3-25 shows a block diagram of a system with active compliance.

The second technique, *passive compliance*, provides compliance mechanically by means of a *remote center compliance* (RCC) device. The concept was originally developed in the 1930s to reduce vibrational stresses in radial aircraft engines by permitting deflections in the engine mounts which coincided with rotations about the engine's center of gravity. The theory of operation of RCC devices is a little abstract, but understanding the principle of operation is a minimum requirement for anyone using these devices in a work cell. The *center of compliance* (remember, compliance implies movement) is that point at which the entire compliance system is considered to be concentrated and acting. The location of the center of compliance is determined by the design of the device, and depends on the location of the

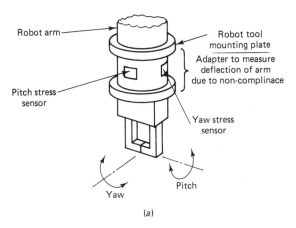

Figure 3-25a Active Compliance Adapter

Sec. 3-5   Assembly Fixtures   83

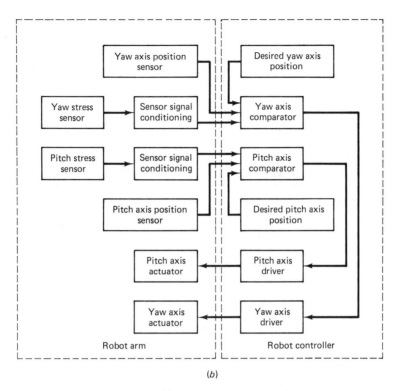

Figure 3-25b Block Diagram for Active Compliance

compliance elements and their orientation. As the name indicates, the center can be relatively remote from the device itself. Figure 3-26 illustrates this concept. The elements consist of four rods for lateral compliance (only two rods are shown in Figure 3-26) and four rods at an angle for rotational compliance (again, only two rods are shown). This device is rigid in the axial direction with no compliance provided. This particular design is a result of research work at the Charles Stark Draper Laboratory at MIT. The operation of the RCC device in lateral and rotational compliance is illustrated in Figure 3-27. Note that the RCC is mounted between the gripper and the robot tool plate or wrist, but the required motion for alignment results from forces applied to the part in the gripper. In Figure 3-27a the force on the pin which results from an offset from the hole causes a lateral movement in the pin. The pin remains in line with the hole during the lateral shift because of the RCC device movement. In a similar manner the pin in Figure 3-27b is rotated into alignment with the hole as a result of the force applied by the hole and the action of the RCC elements. The outer elements provide lateral compliance, and the inside elements produce the rotational compliance (see Figure 3-29).

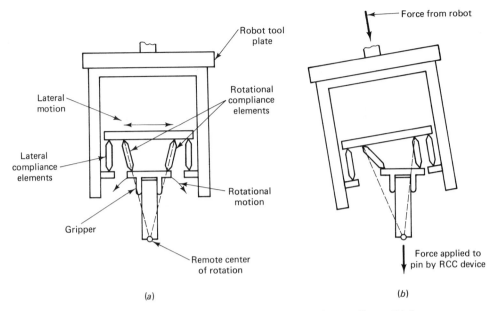

Figure 3-26 RCC with Lateral and Rotational Compliance Links

Another form of passive compliance is found in the SCARA (Selective Compliance Articulated Robot Arm) configuration. The IBM 7535 and 7545 robots in Figure 2-12 use SCARA technology to provide .013 mm of tool movement for insertion compliance at programmed points. After the tool reaches the programmed point, the controller frees the servo-system so the

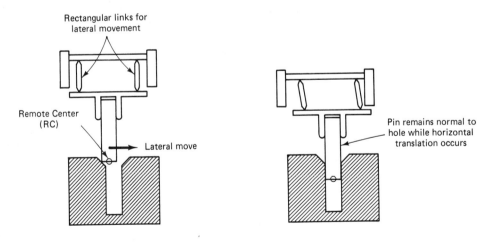

(a) Rectangular elastic linkage

Figure 3-27a Rectangular Elastic Links

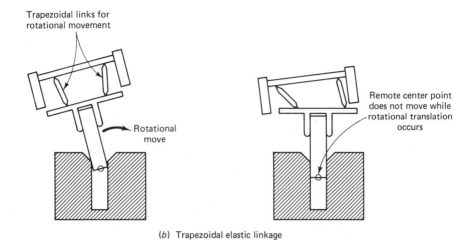

(b) Trapezoidal elastic linkage

Figure 3-27b Trapezoidal Elastic Links

gripper can move the .013 mm. If the gripper attempts to move beyond the permitted range, the servo-system stiffens and the gripper position is maintained. This type of arm is ideal for parts inserted vertically into holes.

Active compliance systems are available from some robot manufacturers, and passive devices are sold in several different sizes by a number of vendors. Sensors are often inserted in the RCC devices to measure the amount of movement in the passive compliance device when the two mating parts are assembled. The sensor data can be used by the host computer to correct the robot sequences for systematic errors introduced by machine aging or drift in calibration settings. The coordinates of the two mating parts can be corrected while the assembly process is operating using the data provided by the RCC sensors.

3-6 MULTIPLE END EFFECTOR SYSTEMS

In most applications currently using robots, the machines have a single function to perform in the work cell, and only one end effector is necessary. However, as work cells become more complex and robots more sophisticated, the application of multiple gripper systems will become more frequent. A *multiple gripper system* is one which has a single robot arm but two or more grippers or end-of-arm tools which can be used interchangeably on the manufacturing process in the cell. These multiple tooling systems can be completely separate grippers or tools mounted to the fixture on the end of the robot arm. In other applications the robot tooling is designed to permit two or more grippers to be used on the same robot arm on an interchangeable basis. In many work cells the robot automatically changes to the correct tool during the production process.

As work cells become more complex and costly, the concept of increased flexibility through multiple end-of-arm tooling makes good economic sense. Cost justification on this basis alone is difficult to establish in terms of man-hours saved, but in many production situations the use of robot automation would not be possible if multiple grippers were not employed. For example, in a production situation where the range of part sizes is great, the only way to avoid installing several robots is through multiple grippers. Or in an assembly application where the number and variety of operations to be performed is numerous, it would not be economically feasible to have a robot and separate gripper for each operation.

The advantages of using multiple grippers include (1) increasing the production capability of the work cell, (2) reducing work-in-process time for the part, since it must be moved through fewer work stations, and (3) utilizing the robot arm and controls more efficiently. In addition, many assembly applications would not be possible without the development of multi-gripper work cells. The primary disadvantage is the additional complexity which is added to the tool design problem. This also results in greater cost for the design and construction of the devices.

The multiple gripper system pictured in Figure 3-28 is used in a work cell developed by Fared Robot Systems which permits assembly systems to be built and tested in a development laboratory.

Figure 3-28 Multiple Gripper System (*Courtesy of Fared Robot Systems*)

Figure 3-29 RCC Devices from Lord Corporation (*Courtesy of Lord Corporation*)

3-7 VISION AND ARTIFICIAL SKIN

Future robot systems placed in complex manufacturing cells will have the ability to see and feel the parts manipulated in the process. Since there is no reason to require the eyes or vision system cameras of these robots to be mounted above the robot or, in effect, in the robot's head, the camera could be placed on the arm above the gripper to get a close look at the work to be done. It could even be placed in the end-of-arm tooling to look directly at the work. Also, the eyes of future robots need not be human-type vision systems which see in color and three dimensions. They may be a simple *photocell* to detect part presence or a *laser system* to locate the seam for a welding-tracking system. The eyes of the robot could be located in the palm of the gripper and need only supply the information which the system requires to produce the product.

The ability of a robot to recognize the presence and quality of a part through tactile sensing is necessary for many manufacturing tasks but currently exists only as a research and development activity in university and industrial laboratories. The result of the research will be an *artificial skin* which can sense the size, weight, and texture of the part as it is grasped by the gripper.

3-8 SUMMARY

The objective of this chapter was to introduce you to the different types of tooling which are currently available for use on robot arms. Grippers must be capable of holding the required part so that safety is assured under all operating conditions. In addition, they should conform to the lightest design using the simplest technique possible while holding and sensing the presence of the manufactured part.

In general, end-of-arm tooling groups include standard grippers, vacuum grippers, special-purpose grippers, and special-purpose tools. Assembly fixtures require that special attention be paid to the problem of compliance.

The alignment problem between mating parts can be solved either by active or by passive compliance. In active compliance the compensation for part alignment error takes place in the commands the controller sends to the arm. The adjusted coordinate values are generated as a result of signals received from the gripper indicating a binding condition in the mating parts. The second type of compliance is called passive and uses a remote center compliance (RCC) device. The required movement of the part to permit alignment and mating is produced in the RCC, which is mounted between the gripper and the robot wrist-mounting plate. The corrective movement results from forces acting on the part in the gripper as it is inserted into the assembly without perfect alignment.

Multiple end effector systems have the advantage of permitting robot automation to be implemented in work cells involving complex operations. In addition, they permit increased utilization of expensive robots on a wider range of manufacturing problems. These advantages overcome the primary disadvantage, which is increased cost because of greater complexity.

The rate at which robots are introduced into new production situations is directly related to the rate at which the tooling develops. Many of the new applications require grippers that approximate the dexterity of the human hand. The development of a gripper with three or more fingers which can grasp a variety of irregularly shaped objects is critical for many future applications. Increased robot utilization is directly proportional to technical breakthroughs in end-of-arm tooling design.

QUESTIONS

1. What is end-of-arm tooling and what function does it serve?
2. What are the five characteristics which all end-of-arm tooling must satisfy?
3. What is conservation of job IQ?
4. How can the IQ of the job be used in the selection of a robot system?
5. How can the IQ of the job be used to reduce the cost of a robot application?
6. What are the three categories for classifying end effectors, and what is included in each category?
7. Draw a classification tree which includes all the types of end-of-arm tooling described in Section 3-3.
8. What is the estimated cost for off-the-shelf and in-house designed tooling as a percentage of total system cost?
9. Write a procedure for calculating the lifting power of round vacuum cup grippers.
10. What does the term *compliance* mean?
11. Describe the two methods used to achieve compliance.
12. Describe the three compliance conditions normally found in robot applications.

13. What is an RCC?
14. What are the advantages and disadvantages of multiple end effector systems?
15. What will artificial skin allow the next generation of robots to do?

PROJECTS

1. Design a decision tree which will select the best end-of-arm tooling for a manufacturing problem. Use questions at each branch of the tree that require a *yes* or *no* response.
2. Write a computer program which will execute the decision tree from Project 1.
3. Write a computer program which will input the size and number of vacuum cups in a gripper along with the degree of vacuum achieved and output the lifting capability of the gripper.

4

ROBOT AND CONTROLLER OPERATION

4-1 INTRODUCTION

A study of robots and controllers must start with a definition of *servomechanisms* which are the basis for control in all high-technology robot systems. A *servo system*, as it is frequently called, consists of a high-power device or drive motor, a low-power control mechanism, and an interface electronic system which moves the high-power device or drive motor under the control of the low-power mechanism. The electronic interface amplifies the small control signals in order to operate the high-power drive and move it in the desired direction. The position of the high-power drive output shaft is constantly monitored by the electronic system with sensors, so that the difference between the desired shaft position and the actual position of the drive approaches zero. Figure 4-1 is a block diagram of a simple servo system.

Current industrial robots can be divided into *servo type* and *non-servo type*, depending on the control technique used. In non-servo or open-loop machines the controller does not use feedback sensors to accurately control drive power to the arm. The servo-driven robots use three different types of feedback sensors to close the loop and determine arm position. The three types include: *potentiometers*, which use resistance to measure the position of the robot joints; *optical encoders*, which use photo-optics for measurement; and *resolvers*, which use magnetics. Using feedback sensors, the position of the robot tool-center-point can then be calculated within a reference coordinate frame.

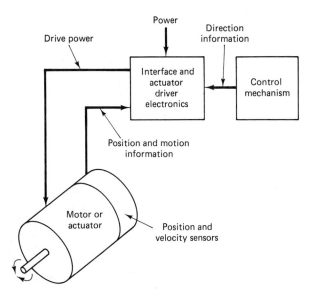

Figure 4-1 Servo System Block Diagram

4-2 REFERENCE FRAMES

Reference frames are sets of Cartesian coordinates which describe the positional relationship between the robot, the gripper, the tool-center-point, the workpiece, and the universe in which all of these exist. The dimensional relationships between the parts of the robot and the production parts in the work cell are described by means of mathematical equations called *transformations*. Figure 4-2 illustrates an idealized robot located in a universe frame along with the workpiece, gripper, and tool-center-point. The robot controller can calculate the necessary joint angles that will allow the tool center reference frame to be aligned with the part or workpiece reference frame. The program stored in the controller performs the calculations using the transformation equations developed for the robot by the robot manufacturer. These frames and transformations are completely transparent to the robot programmer or operator; however, they are used by the robot designer who is developing the robot operating system and control algorithms. Although space does not permit a complete development of this concept, a basic understanding of the process is useful to the robot technician and applications person.

The position, or angle of every robot servo-driven joint, is known from the sensors which feed back that information to the robot controller. The location of the gripper with respect to the robot frame is not known from any single feedback sensor but must be calculated by the robot controller

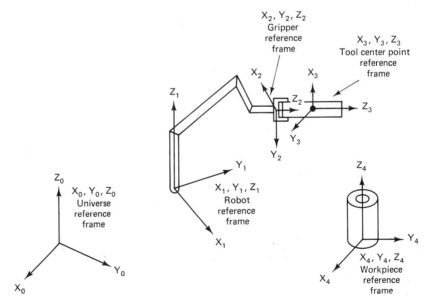

Figure 4-2 Reference Frames

Sec. 4-2 Reference Frames

from the data received from all the joint sensors. For example, what is the value in *X*, *Y*, and *Z* coordinate units for the gripper in the robot reference frame in Figure 4-2? That position can be calculated if equations are written to describe the relationship between the gripper reference frame and the robot reference frame. The equations would have as variables the joint angles of the robot arm so that as the joint angles changed, the displacement of the gripper from the *X*, *Y*, and *Z* axes could be calculated. In addition, the equations would contain some constants, such as the length of each arm element and the distance from the tool plate to the grip point of the end-of-arm tooling. Similarly, if the equations which relate the robot reference frame to the universe reference frame were known, the position and orientation of the gripper in the universe reference frame could then be calculated. Figure 4-3 shows a robot in a work cell with all frames identified. Since the relationship of the frames to each other is known and described by equations inside the robot controller, the displacement of the part and the

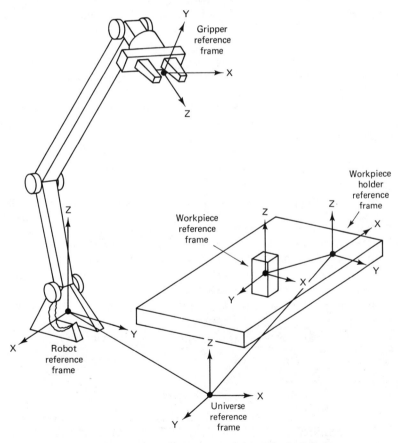

Figure 4-3 Reference Frames in a Work Cell

gripper from the universe reference frame axes can be calculated by the controller. Thus, in order for the robot to grasp the part, the controller need only change the joint angles, using the equations, until the gripper reference frame and part reference frame have the same displacement and orientation from the universe reference frame. Under these conditions (see Figure 4-4), the frame of the gripper and the frame of the part are aligned. This alignment would not have been possible without the equations that describe the relationship between the frames in the system.

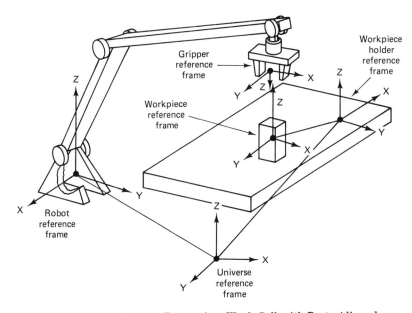

Figure 4-4 Reference Frames in a Work Cell with Parts Aligned

The feedback sensors on each axis of the robot also provide the controller with data on the velocity and acceleration/deceleration of each joint as it is moving. These data are used to calculate the velocity and acceleration/deceleration of the tool-center-point. The controller is responsible for the rate of change in the tool-center-point between two programmed points, and it sets the rate of change of each joint angle so that the correct tool velocity and acceleration/deceleration is achieved.

4-3 OPEN-LOOP SYSTEMS

Non-servo, or *open-loop* systems, account for over one-half of the current robot applications in the United States and nearly 65 percent of those abroad. The advantages of an open-loop system include:

- Lower initial investment in robot hardware
- Well-established controller technology in the form of programmable controllers
- Less sophisticated mechanical and electronic systems with fewer maintenance and service requirements
- A larger pool of engineers and technicians familiar with the requirements of this type of system

Internal Control

The control of non-servo robots includes internal mechanisms for establishing accurate positioning and external drive devices for ensuring the required sequence of movements. *Internal positioning mechanisms* include the following:

- *Fixed hard stops* which limit the movement of pneumatic or hydraulic actuators at each end of their travel. These stops can either coincide with the natural extension length of the actuator or can be added to the actuator to limit travel.
- *Adjustable hard stops* which limit the movement of pneumatic or hydraulic actuators at each end of their travel. These are identical to the fixed hard stops except that they are adjustable over part of the length of the actuator extension. Figure 4-5 shows the adjustable stops on the Y transporter of a Mack modular robot unit. Figure 4-6 illustrates adjustable stops on an IBM 7535 robot system. Note that the stops can be adjusted under program control using pneumatic cylinders.

Figure 4-5 Adjustable Hard Stops on B-A-S-E Robot (*Courtesy of Mack Corporation*)

Figure 4-6 Limit Switches for Variable Stops (*Courtesy of Fared Robot Systems, Inc.*)

- *Limit switches* which produce variable stops or extensions on non-servo actuators. The Prab robot pictured in Figure 2-26 shows the limit switches used to produce variable degrees of rotation of the arm on the base.
- *Stepper motors* which produce a fixed degree of rotation based on the number of pulses which are applied to the motor windings. Stepper motors are not often used on industrial robots but are common in peripheral hardware, such as positioning tables.

The rugged nature of the internal position control devices assures long operating life and high mean-time-between-failure (MTBF). The pneumatic systems require only clean dry air, line filters, and an oiler to provide trouble-free performance. Some pneumatic systems from Schrader Bellows are pre-lubricated for life so that oilers are not necessary.

Of course, hydraulic systems require additional maintenance as a result of the pump and tank which is a part of all high-pressure oil systems. With regular filter changes and chemical tests of the hydraulic fluid, however, the non-servo hydraulic robots provide a good MTBF.

External Control

The sequential movements of non-servo robots are controlled by devices external to the robot's arms. The types of controllers used by pick-and-place robots include:

- *Drum programmers*, such as the Seiko sequencer illustrated in Figure 4-7, are used in some applications. In this unit, as in most mechanical

Sec. 4-3 Open-Loop Systems

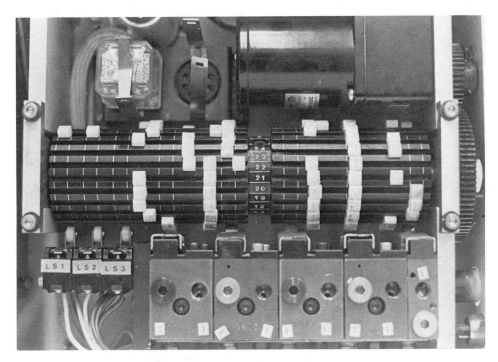

Figure 4-7 Drum Programmer for Seiko 700 Robot

drum programmers, the operational sequence is set by the arrangement of tabs or cams on the drum surface. The tabs actuate either electrical switches or hydraulic/pneumatic valves which control the movement of each robot axis. The time for each sequence is determined by the number of tabs used and the speed of rotation of the drum.

- *Pneumatic logic* or *air logic programmers* are also used to control the sequences of pick-and-place pneumatic robots. An air-powered logic network is built using fluidic logic elements to provide the sequential control the robot requires. The air logic system is built and programmed by connecting the fluidic elements together by small plastic air lines. The time and sequence of robot movements is determined by the fluidic elements used and the way in which they are interconnected.
- *Programmable controllers* (PCs) are the most frequently used devices to control the sequential movement of pick-and-place machines. A programmable controller is an off-the-shelf computer-based industrial controller which is used in all types of industrial applications. Figure 4-8 illustrates the use of a PC in a typical low-technology robot system. Note the use of the PC not only to control the robot operation but also to monitor sensors and control the system pilot lamps.

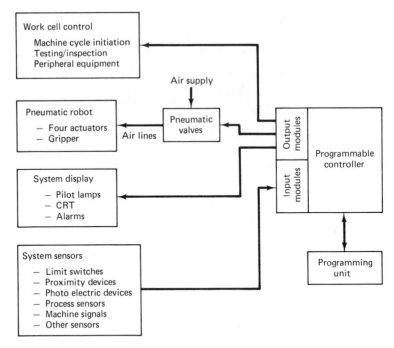

Figure 4-8 Programmable Controller Driven Robot System

Operation

The operation of controllers for non-servo robot systems varies with system size and control technique used. For example, if a drum controller is used the operator would just turn power on after the tabs are placed in the correct position on the drum. In comparison, a system which uses a programmable controller to operate the robot and other work cell hardware may require a greater amount of operator involvement. The program may have to be loaded into the PC from magnetic tape, and the operator may have to interact with the production process performed in the work cell.

4-4 CLOSED-LOOP SYSTEMS

The entire group of robots classified as high technology has servo control systems. This type of system requires a controller which must calculate the current position of each joint and an arm with sensors to provide the data which the controller requires for these calculations. The result is a much more complex controller with a statistically higher chance to malfunction. However, the advantages gained in the application of servo-controlled robots to certain manufacturing processes outweigh the problems created by the additional complexity inherent in the system. These advantages include:

- Servo-controlled robots provide *highly repeatable positioning* anywhere inside the work envelope. This flexible multiple positioning characteristic is necessary for the implementation of robots in many production jobs.
- The servo-type controller with computer capability can provide *system control* for devices and machines which are external to the robot system. Many work cells require the harmonious control of numerous peripheral devices, such as NC machines, conveyors, gauges, sensors, and readouts. The servo-type controller usually has the I/O capability and the programming power to control the robot arm and the other support equipment in the work cell.
- Powerful *programming commands* are available to perform complex manufacturing tasks with very simple program steps. For example, palletizing any size part on any size pallet takes only a few commands.
- *Interfacing* robots to other computer-controlled systems, such as vision and host computers, is easier to accomplish with a servo-type controller.

Internal Control

As in the non-servo systems, the position and motion of the gripper of a servo-type robot is controlled from sources both internal and external to the robot arm. Internal control sources include the *feedback devices* which mechanically measure the angle of the joints and convert the angular measurements to proportional electrical signals. In addition, other feedback devices provide rate-of-change data to be used in controlling the velocity and acceleration/deceleration rates at the tool-center-point.

Potentiometers

The three primary *positional* feedback devices are the *potentiometer*, the *resolver*, and the *optical encoder*. The potentiometer, or pot, is a variable resistor with a linear resistance and a movable wiper. The resolver uses magnetic coupling between transformers to measure rotation, and the encoders use an interrupted beam of light to determine position. In every case the device is attached to the axis drive motor either directly or through reduction gearing. The rotational displacement of the drive motor is measured by the positional feedback device. The potentiometer is usually a single-turn type with gearing used to reduce the multiple turns of the motor to a single turn of the potentiometer. The gearing is either provided by the robot manufacturer or is placed inside the potentiometer by its manufacturer, similar to standard ten-turn potentiometers. Figure 4-9 illustrates a simplified system with a dc servo, arm drive shaft, potentiometer gearing, and potentiometer. Assuming the arm drive rotates through 180 degrees for three turns

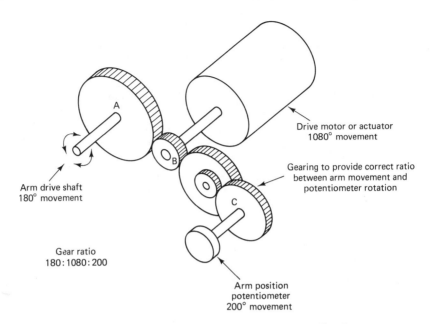

Figure 4-9 Position Feedback Potentiometer Gearing

of the motor and the potentiometer has a rotation of 200 degrees, then a relationship exists between position of the arm and that of the wiper on the potentiometer. Zero degrees on the arm might equate to zero degrees on the pot so that the wiper would be in position A in Figure 4-10. The pot output would be zero volts as a result of the wiper connected to ground. If the robot arm moves to the 180-degree point, then the pot wiper is in position B, and the output might be 4.3 volts. Without any error in the gears or potentiometer, each degree of arm movement would cause the following change in output voltage:

$$\text{Change in V output per degree} = \frac{\text{Change in output voltage in volts}}{\text{Change in shaft rotation in degrees}}$$

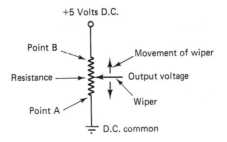

Figure 4-10 Potentiometer

Even with the best resolution and accuracy possible in commercial potentiometers, the error introduced by the pot makes it difficult to use as the primary position feedback device. Instead, potentiometers are sometimes used for coarse position data in systems which combine potentiometers and optical encoders. In addition to lack of accuracy, the potentiometer suffers from the following shortcomings:

- Movement of the wiper contact on the resistance causes wear, which will eventually result in system failure.
- Potentiometer output is affected by the environment.
- The analog output requires analog-to-digital conversion electronics to attain the necessary digital output for the controller computer.

Optical Encoders

Absolute and *incremental* types of *optical encoders* are frequently used as the primary feedback device to measure robot joint movement. In both types, Figure 4-11, a beam of light is interrupted by transparent marks on a rotating opaque disk. The disk can also be glass with opaque marks produced by depositing metal in a precise pattern. The light source is either a light-emitting diode (LED) or a neon or tungsten lamp which produces greater signal strength as a result of greater light output. The receiver is usually a photodiode or phototransistor, but photocells have been used in some encoders. In Figure 4-12 an incremental type of encoder is drawn. Note

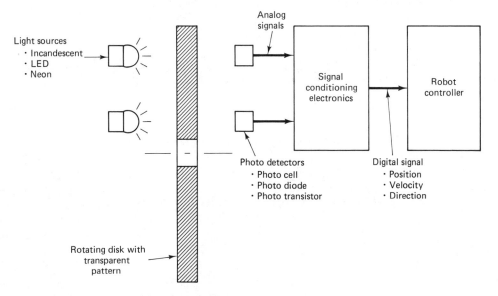

Figure 4-11 Optical Encoder

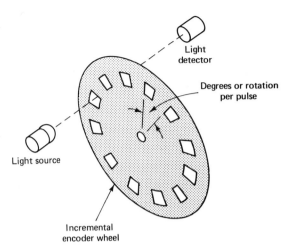

Figure 4-12 Incremental Encoder

that the disk is opaque, and the clear marks are evenly spaced around the entire outer edge. The LED source is separated from the receiver by the disk, and the beam is focused on the edge where it can be interrupted by the marks. This encoder is called incremental because the signal it produces is a series of pulses from the receiver as the light beam is interrupted. Each pulse represents a number of degrees or the increment of rotation necessary to produce it. The typical pulse output waveforms from this type encoder are illustrated in Figure 4-13. The waveform in Figure 4-13a is the actual output from the receiver on the encoder, and the waveform in part b of the figure is the pulse train generated by the encoder electronics. Each time

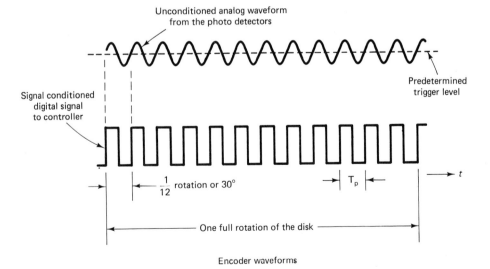

Encoder waveforms

Figure 4-13 Encoder Waveforms

the raw pulse crosses a predetermined value, a single square pulse is produced electronically. As a result of this conversion process, the optical encoder actually functions as an opto-electromechanical sensor to monitor rotational position.

The actual rotational position of the encoder shaft is not known; only the number of increments changed from the previous position is provided. Nor does the encoder in Figure 4-12 provide direction information, since the same pulse train will be produced regardless of the direction of rotation. Direction information is included when a second light beam is added to the system. Figure 4-14 shows an incremental encoder with direction sensing added. The two rows of marks are slightly offset so that for counterclockwise rotation the first beam is interrupted before the second beam is broken by the bottom slot. When the disk is rotating clockwise, beam two is broken first, and then beam one. A digital circuit determines which beam is leading the other and then the direction of rotation is revealed. Using the time T_p in Figure 4-13, the controller is provided the rate of change to calculate velocity and acceleration/deceleration data. Note also that the resolution or accuracy of the encoder is directly proportional to the number of marks on the disk, which corresponds to the number of pulses per revolution. For a resolution of 0.1 degrees the disk would need 3600 marks.

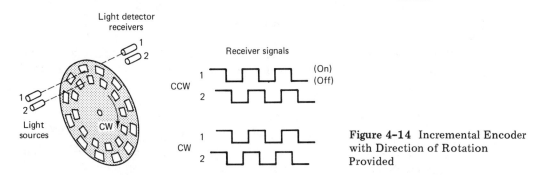

Figure 4-14 Incremental Encoder with Direction of Rotation Provided

An *absolute-type* optical encoder eliminates the primary shortcoming of the incremental system by providing actual rotational position information. This is accomplished by producing a multi-digit binary code for every incremental change in the disk position. Figure 4-15 shows a four-bit absolute encoder. Note that each bit requires its own ring of transparent patterns and a separate light source and receiver for every bit of code or track on the disk. The resolution on the four-bit encoder is very bad since there are only 16 possible codes or different positions of the encoder disk. The resolution in degrees for 16 positions is:

$$\frac{360 \text{ degrees}}{16 \text{ different codes}} = \frac{22.5 \text{ degrees}}{\text{code}}$$

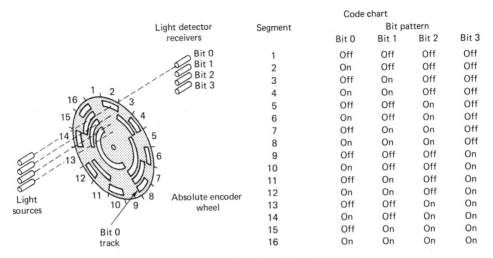

Figure 4-15 Absolute Encoder with Code

Detection of a small rotational change will require a disk with many more tracks and a large number of bits in the code. The code produced by the disk in Figure 4-15 is outlined in the table in Figure 4-15. The code on/off pattern indicates the rotational position of the shaft, and the direction is determined by the increasing or decreasing order of the patterns generated.

Both types of optical encoders are used in robot control systems because they have a number of advantages over the potentiometer described earlier. Their advantages include

- Greater resolution and accuracy
- Noncontact measurements which reduce wear and improve reliability
- No loss of resolution in the conversion of encoder output to controller-compatible code (e.g., a ten-bit absolute encoder will have a resolution of ten bits)

The two disadvantages encoders have when compared to potentiometer-type positional measuring sensors are higher cost and larger physical size, especially with absolute encoders with high resolution.

Resolvers and Synchros

A *synchro* is another type of device used for internal control of servo-driven robot arms. The word *synchro* is a generic term covering a range of ac electromechanical devices which are basically variable transformers. Figure 4-16 shows the schematic diagram and wiring of the two basic types frequently used. The synchro has three windings mounted at 120-degree intervals

Sec. 4-4 Closed-Loop Systems   107

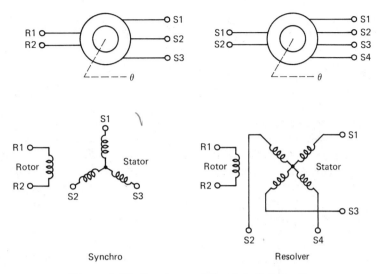

Figure 4-16 Synchro and Resolver Schematics

around a stationary stator, while the *resolver* has two windings offset by 90 degrees around its stationary stator. Both types have either a single or a double winding, called the rotor, which rotates inside the stator windings. The degree of magnetic coupling between the rotor and stator windings varies with the shaft angle or amount of rotation present in the rotor. With an ac reference voltage applied to the rotor winding, the stator windings have an ac output which is a function of the coupling coefficient or shaft angle of the rotor. The amplitude of the ac output present on the stator windings will vary as a function of the shaft angle and will be represented by a sine wave on one output and a cosine wave on the other. In order to interface the output of the resolver with a digital-based robot controller, the analog position data must be converted into binary code. The three widely used methods of conversion are (1) time-phase shift, (2) sample and hold, and (3) tracking. The basic block diagram for a resolver-based position encoder is given in Figure 4–17.

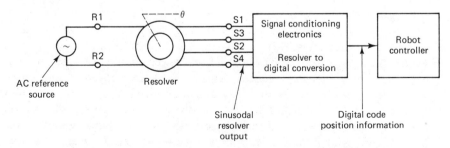

Figure 4-17 Block Diagram of a Resolver Based Position System

A comparison of the resolver-based position encoder to the optical encoder results in the following advantages for the resolver-based system:

- The optical encoder is more dependent upon mechanical precision to achieve positional accuracy while the resolver relies more on electronic circuits for the same accuracy.
- In the resolver the conversion electronics are relatively remote from the resolver unit itself. Thus the electronics unit can receive greater protection from environmental conditions. In the optical encoder, the conversion electronics are in the unit because signal strength from the optical unit is low.
- The resolver-based system is less susceptible to damage since there are no glass disks or light sources which can burn out.
- The resolver has smaller physical size at higher resolution values.
- The resolver requires fewer output wires between the sensor and controller for absolute encoding.
- The position sensing of the resolver is always absolute.

The resolver-based system does have some disadvantages when compared to the optical encoders.

- It is more expensive than an incremental encoded system, especially at higher resolutions.
- It needs an ac reference voltage.
- It requires a conversion unit which has one to two bits of resolution greater than that needed by the system as a whole; this results from the resolver-based position system's having an accuracy which is one or two bits less than the stated resolution.

Most servo-controlled robot systems use either optical or resolver-based positional encoders to determine the present joint angles for use in controller calculations. However, the popular Puma robot from Westinghouse/Unimation uses a combination of potentiometers and optical encoders for position information. Potentiometers are used for coarse location of the joints, and incremental encoders are used for fine positioning data. In general, optical encoders have greater application at lower resolution values, and resolvers are used when greater accuracy and higher resolution are necessary.

The last type of sensor used to measure linear position of robot axes is illustrated in Figure 4-18. The system uses sonic pulses which are generated at the reading head by a technique called *magnetostriction*. The pulses travel along a specially heat-treated nickel-iron wire buried within the long measur-

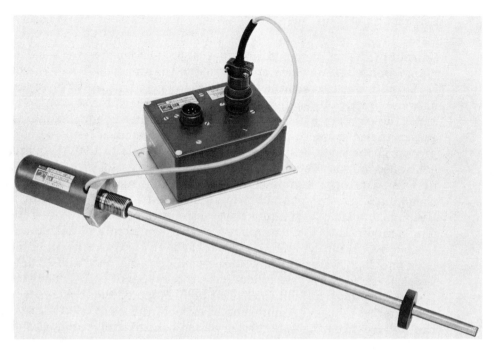

Figure 4-18 Sonic Displacement Transducer (*Courtesy of Temposonics, Inc.*)

ing spar of the metering unit. The displacement value is based on measuring the time interval required for a sonic pulse to travel between two points. The sonic pulse is launched at the reading head by the interaction of a current pulse applied to the wire and a magnetic field. The torsional stress wave that results travels along the wire at a rate equal to 0.11 inches per microsecond. At the reference end of the spar, a transducer detects the pulse, and a signal to determine the time interval between launch and arrival of the sonic pulse is generated. The unit pictured in Figure 4-18 is supplied by Temposonics Inc. and measures the displacement of the toroidal magnet along the spar. Ultrasonic displacement transducers have the following advantages:

- Non-contact measurement of the moving axis which produces no wear in the transducer system
- High reliability due to no moving parts
- Infinite resolution in the analog mode with high linearity and excellent repeatability
- Direct digital output of displacement data

External Control

Control of the programmed motion in a servo-driven robot arm is dependent upon several factors which are external to the arm. The first of these factors is the *path-control technique* which the controller uses to drive the robot between programmed points. A second factor, which is also common to controller design, is the *programming language*. Some robot programming languages utilize commands that alter arm motion as a result of conditions present in the work cell, a factor which is external to both the arm and the controller. Interactive control of the robot arm is achieved on some models by using data from sensors in the work cell. The signal from a remote sensor is fed back to the robot controller and used to change the arm motion. Vision is an example of interactive control in which the motion of the robot arm is modified so that the robot can grip a randomly placed part which is located by a vision system.

The first factor, path-control technique, can be classified into three types. In servo-controlled robots these three types of path control are *point-to-point*, *continuous*, and *controlled path*. None of the three types relies on any external signals or commands to execute the resultant motion. The motion follows from the basic design of the controller and cannot be changed. A continuous-path robot will always be programmed as a continuous-path machine and will always execute the programs in that fashion. The arm is controlled by the motion control technique built into the controller. Each of these path-control techniques was described in Chapter 2.

The second factor in external control of the arm is the programming language. The controller drives the arm through a sequence of moves based on a program stored in the controller memory. This program is written by the robot programmer using the language built into the controller. On some robots the language is synonymous with the motion and programming buttons present on the teach pendant, while on other systems the language consists of the teach pendant positioning plus those system commands which are integrated into the programmed motion from a keyboard. Branching is an example of a program command entered from the keyboard which can alter the motion of the arm. When the robot executes a program step which has a branching command, the controller has two program paths that can be followed, depending on the value of the branch variable. For example, in the Cincinnati Milacron controller the command

$$\text{PERFORM 20, +S5}$$

directs the robot to follow the program steps in Sequence 20 if input 5 is turned "on." If input 5 is turned "off" the branch to 20 is not executed, and the program continues in the current sequence and executes the program step which follows. Each of the two program directions would produce different arm motion; in this way, the program or external input has control of the arm movement.

The last type of external control for servo-driven robots, that of feedback from *external sensors*, puts the arm into a real-time interactive control mode. Real-time interactive control means that robot arm motion is altered as it moves through a programmed work exercise. For example, an ASEA robot which is programmed to grind the burrs from castings will have its motion modified by the size of burrs that the grinding wheel encounters. Because the size and location of a burr is completely random, the robot must adjust in present or real time, as it is called, to the burrs encountered. A sensor determines the size of the burr which the wheel is grinding and sends a feedback signal to the controller to modify the program. This modification causes the robot to grind longer on the larger burr.

Another type of real-time interactive control is seam tracking in MIG and TIG welding, where the seam to be welded is located by a photo sensor, a vision camera, or by welding parameters. The location of the seam or welding path is passed to the robot controller, which moves the arm holding the welding tooling. The location and control is dynamic, so that corrections in the direction of the welding gun are made as the weld is in progress. Another example is the tracking of moving parts on conveyors. The robot arm compensates for changes in conveyor speed as it works on parts passing on a moving assembly line. The most dramatic example of real-time interactive control is the use of a vision system to locate and identify randomly placed parts. The interaction occurs when the vision system passes the location and orientation information to the robot controller so that the robot arm can position the tooling to grip the part.

Operation

Operation of servo-driven robots is often more difficult than that of their non-servo counterparts. System start-up involves a series of keystrokes from the keyboard and/or teach pendant which is followed by the teaching process to satisfy the application. After testing and debugging the application program in the controller memory, the program is usually saved on disk or magnetic tape for backup. The program will remain in the robot controller memory as long as the main power is not interrupted and the controller backup battery is charged. The system is ready to execute the current program stored in memory whenever the robot is powered up.

4-5 CONTROLLER ARCHITECTURE

The block diagram in Figure 4-19 illustrates the typical modules found in a controller for a servo-controlled arm. In Figure 4-20 the architecture for a non-servo machine is shown. The heart of both systems is the *Central*

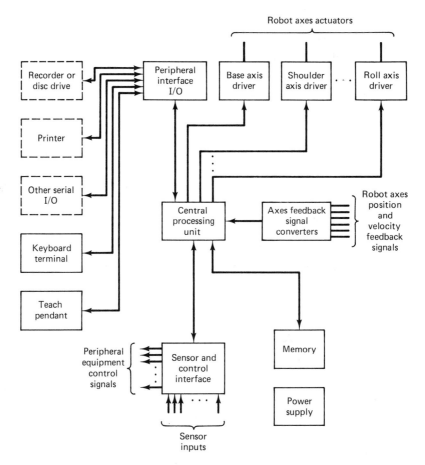

Figure 4-19 Controller Block Diagram for a Servo System

Processing Unit (CPU) which is responsible for memory management, input/output (I/O) management, information processing, computation, and control of each robot axis. The CPU configurations for current robots demonstrate the alternatives provided by the explosive growth of computing hardware. The options adopted by some robot manufacturers include:

- Using a minicomputer to handle all of the responsibilities assigned to the CPU. Cincinnati Milacron uses an in-house designed minicomputer for the controllers of the popular T3 hydraulic series of robots. IBM uses a Series 1 model from their line of computers to provide the control functions on the model 7565 manufacturing system.
- Using off-the-shelf single board computers, such as the Digital Equip-

Sec. 4-5 Controller Architecture   113

ment Corporation LSI-11, to configure a special-purpose computer capable of handling robot CPU duties. The PUMA series of robots from Westinghouse/Unimation uses this technique in controller design.

- Using either an 8- or a 16-bit microprocessor as the base of the CPU design. Many of the Japanese manufacturers have adopted this technique, and also the manufacturers of programmable controllers use a single microprocessor in most cases.
- Using a separate microcomputer interfaced to the robot controller to provide the command programming and computational power needed for arm control. IBM uses their PC with the 7535 and 7545 model robot arms, and the Rhino RX-1 from Rhino Robots will interface to any computer through the robot controller's RS232 serial port.

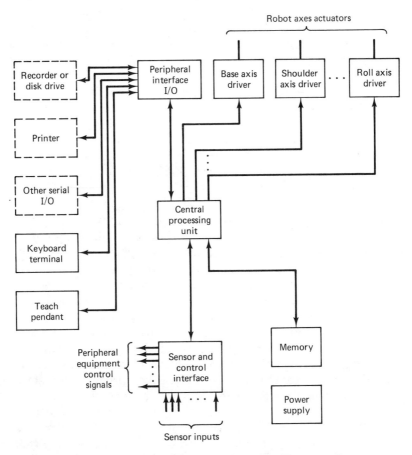

Figure 4-20 Controller Block Diagram for a Non-servo System

- Using a network of 8- and 16-bit microprocessors linked together by hardware and software to perform the function of the CPU. The new Acramatic Version 4.0 robot controller introduced by Cincinnati Milacron for the electric robot models uses this type of architecture.

This last technique described, networking, offers a number of advantages for robots in the high-technology group. The system operation is much faster when the duties of the CPU are divided up among different processors. With complex arm control algorithms, the processing speed within the CPU becomes critical; either faster processors or multiple processors are required.

The memory used in most robot controllers is solid-state with battery backup for program protection during loss of primary ac power. The operating system and application program are usually stored in volatile devices (RAM), and nonvolatile devices (ROM) are used to store basic load routines to wake the system up and load the controller programming language. The primary exception to this occurs in pick-and-place robots, which use mechanical programming devices for program storage. An example of this is the drum programmer, which uses the position of stops on the drum to actuate the robot axes.

The CPU controls the interface between two different types of peripheral hardware. The first is *digital data communications* between the controller and the peripheral devices needed to operate the robot system. The most commonly used peripherals are the teach pendant and the keyboard terminal. More complex systems often include a recording device for program storage and playback, a printer for hard copy, and a general-purpose serial port for communications with other computer-controlled hardware such as a host computer or vision system. The second type of interface is for *discrete* or *on/off type signals*. This interface in the servo-type robot controller monitors the output of sensors in the work cell to determine if a change of state has occurred. For example, a sensor to detect the presence of a part in the gripper would be monitored through an input of this interface. If the gripper acquired the part, the sensor output would change from *off* to *on*, and the controller would drive the arm to the next programmed point only when this input change was received through an input port of the interface I/O. The output portion of the interface has the capability to control discrete devices. Under program control, devices in the work cell can be turned *on* or *off* through this output section of the controller. For example, in a punch press loading application the robot program can activate the press to start processing a part which has just been loaded.

The discrete interface section of the robot controller is basically a built-in programmable controller for use in the work cell. Typical input values of 80 to 140 volts ac and 6 to 40 volts dc are available on the discrete input interface of some robot controllers. The output interfaces will switch either ac or dc currents in ranges that are compatible with the input section. In some applications the number of discrete I/O points which must be monitored exceeds the I/O capacity of the robot controller. When this occurs, a standard industrial programmable controller (PC) is used to handle the excess control requirements. The external PC is interfaced to the robot controller through an input on the controller discrete I/O. Figure 4-21 shows a standard PC monitoring work cell sensors and passing their status on to the robot controller through a single input channel. For example, the robot could require that five sensors be activated before the next move can be executed. To use the robot input interface would require five input channels, but if an external PC is used to determine the status of the sensors, then only one input is required. The PC monitors the status of the five sensors and passes a *go* or *no-go* signal to the robot.

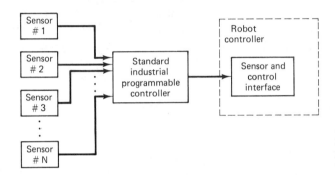

Figure 4-21 Block Diagram of a Robot Controller and External Programmable Controller

The capability of PCs for robot control is increasing with every new model or option which is released. In many robot applications the robot controller is only a standard industrial PC. A pick-and-place robot needs only an *on* or *off* signal to the axis control valve to drive the axis between the two limits. A PC can provide this type of control and also monitor any sensors which are required by the system. Some new PC models include proportional control modules that permit the PC to be programmed to drive closed-loop servo-type robot arms.

4-6 SUMMARY

The control of closed-loop robot arms requires coordinate systems for every part of the robot and work cell. The relationship between the coordinate systems must be defined through transformation equations, so that the X, Y, and Z position of the tooling can be calculated from the joint position data available in the controller. The joint angles and arm extensions are measured by either potentiometers, optical encoders, resolvers, or sonic displacement transducers. The optical encoders used include both the incremental and absolute types. Most robot systems use either optical encoders or resolvers because the accuracy and reliability of potentiometers is not adequate for industrial applications.

Position sensors are part of the internal mechanisms which control the motion of the robot arm. In pick-and-place machines the limit of motion is set by mechanical stops on the arm. The motion of the arm is also controlled by external factors outside the arm. These factors include the program and operating system stored in the controller and sensor input from the work cell.

Robot controller architecture for servo feedback systems includes a central processing unit (CPU), memory, peripheral I/O, discrete I/O, power converter or supply, axes drivers, and feedback signal conditioning. Although the non-servo robot system also has these elements, the axes drivers are part of the discrete I/O and the feedback network is absent. The non-servo robot controller is often manufactured by the vendor who builds the robot arm, but in some cases a standard industrial programmable controller is used for arm control.

QUESTIONS

1. What is a servo system?
2. What are the four basic sensors used in servo feedback loops, and how do they measure the joint positions?
3. What are reference frames, and how are they used in the control of robot motion?
4. What two parameters are measured at every joint of most servo-controlled robots?
5. What are the advantages associated with non-servo robots?
6. Describe four techniques used on non-servo robot arms to achieve positioning.
7. What are the maintenance requirements on hydraulic and pneumatic non-servo robots?
8. Describe three devices external to the robot arm which produce sequential movement in non-servo robots.
9. What are the operational requirements for non-servo robots?
10. Describe four advantages of servo-controlled robots.
11. The feedback potentiometer on a robot has a voltage change of 4.8 volts with a rotation of 320 degrees. What is the resolution of the system in terms of volts per degree of rotation?
12. Describe four limitations present when potentiometers are used for feedback sensors.
13. Describe how incremental and absolute optical encoders are constructed and how they measure shaft position.
14. Describe how the direction of the rotation is determined in each type of optical encoder.

15. What is the rotational resolution in degrees of an absolute encoder with a 12-bit binary code?
16. How many optical marks will be required on an incremental encoder with a resolution of .05 degrees?
17. What are the advantages and disadvantages of optical encoders compared to potentiometers?
18. What is a synchro?
19. What is the difference between a synchro and a resolver?
20. How do resolvers operate when used as shaft angle encoders?
21. What are the advantages and disadvantages of resolver position detection over optical type encoders?
22. Describe the three factors external to servo robot arms which control arm motion.
23. What are the responsibilities of the CPU in controller architecture?
24. Describe five configurations used in current controllers that fulfill CPU responsibilities.
25. Describe the memory systems used in the controllers.
26. What are the two different types of interfaces handled by the controller CPU?
27. Describe one example of each type of interface.
28. Describe two applications for programmable controllers in production work cells.

5

SENSORS AND INTERFACING

5-1 INTRODUCTION

Even the most unskilled production worker can tell when a part has fallen to the floor or when a finished part is not ejected from a machine. The most sophisticated robot available today cannot perform these routine tasks without the help of sensors. Strategically mounted sensors provide the robot system with the same data which an operator gathers using the five human senses. For example, a sensor would check for the presence of a part in the die of a forming press after the ejector has been actuated. An operator would use vision to determine if the ejector had failed, but the robot system might use a light source and light-sensitive receiver to detect the presence of a part. Figure 5-1 shows this type of sensor detection system when the lower half of a die with part is in place. If the light beam is broken a part is present.

The six basic reasons sensors are used in a work cell are

1. To detect a condition where an operator or some other human worker could be harmed by the robot or other manufacturing equipment
2. To detect a condition where the robot or other machines could be harmed by some other manufacturing equipment
3. To monitor the production system operation to ensure consistent product quality
4. To monitor the work cell operation to detect and analyze system malfunctions
5. To analyze production parts to determine the current level of product quality
6. To monitor production parts for identification, location, and orientation data that can be used by other production systems in the work cell.

Sensors are grouped into three basic categories, called *contact*, *non-contact*, and *process-monitoring* devices. As the name implies, a contact

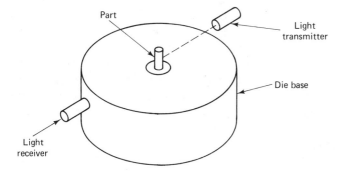

Figure 5-1 Part Ejector Failure Sensing

sensor must physically touch an object before the sensor is activated. With noncontact devices the sensor output changes states before the sensor comes in contact with the object being sensed. Process-monitoring devices include all sensors which measure process parameters, such as temperature and pressure limit switches.

5-2 CONTACT SENSORS

Contact sensors include *limit switches* and all *tactile-sensing* devices. Limit switches have been used in process automation for many years and are today one of the most reliable and easy devices to interface in an automated work cell. Tactile sensing is in the development stage with a high level of research activity in both the industrial and university laboratory.

Limit Switches

The selection and application of limit switches require an understanding of the *physical* and *electrical* properties of the devices along with their *operational characteristics*. As these properties and characteristics are described, frequent reference should be made to the device data sheets included in Appendix A.

The *physical* properties of a limit switch include the fact that it is a mechanically actuated electrical switch consisting of a receptacle, switch body, operating head, and contacting device. On the last page of the Omron Model D4A data sheets (Appendix A) two types of switch bodies, receptacles, and operating heads are shown. Note that the contacting device on one is a lever and on the other it is a roller plunger. The receptacle provides the electrical connections for interfacing, while the switch body holds the switch and provides a mounting base for the various operating heads. The operating heads listed on the last page of the data sheet will mount to a standard switch body. The rotary type requires the contacting devices to rotate to activate the switch, while the two types of plunger heads require a linear movement of the plunger to cause switching of the contacts. The wobble lever type has a long trip rod attached to the head which will trip the switch if the rod is moved away from the center position in any direction. As the data indicate, the different head types are available with various contacting devices. For example, both plain and adjustable plungers are provided, and the wobble lever type has four different contacting rods. Some of the many levers available for rotary type heads include standard-roller, offset-roller, adjustable-roller, rod-adjustable, and fork-roller. Illustrations of the different types of heads and contactors are provided in the data sheets in Appendix A.

The *electrical* properties include the contact ratings for current and voltage and the contact configuration. The contact ratings are broken into two categories, called *pilot duty* and *electronic duty*. Pilot duty assumes switching voltages of 120 to 600 volts ac with currents from 60 to 0.6 amperes, respectively. This would be normal service conditions for controlling motors, motor starters, lamps, and other high-power devices. The ratings table under "Specifications" on the D4A data sheet classifies these currents depending on operating voltage and contact closure (make) or contact opening (break) conditions. Electronic duty covers the operation at low values of voltage, 5 to 30 volts, and low values of current, 0.2 to 100 milliamperes. The measure of a switch's ability to handle low currents is called the *dry circuit rating*. An input module of a programmable controller (PC) requires less than 150 milliamperes when 28 volts is applied. Therefore, limit switches interfaced directly to PCs should have contacts with good low-current switching characteristics. The contact configurations available in the D4A model are illustrated in Figure 5-2. Both are double break type switches, which means the contactor leaves both of the normally closed (NC) contacts when the switch is actuated. In the two-pole model the operation is either sequential or center-neutral type. The sequential type has pole one closing first, followed by pole two as the switch is actuated. In the center-neutral type, pole one operates during clockwise rotation of the lever and pole two during counterclockwise rotation.

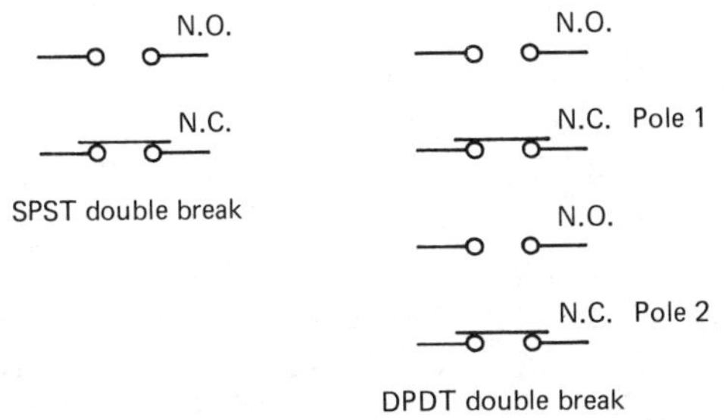

Figure 5-2 Contact Configuration for Limit Switches

The *operational characteristics* of the model D4A switch include everything from the operating speed to the device weight. Review the characteristics listed on the data sheet in Appendix A and note the excellent service life of these devices. The definition of various switch movements and the forces required to produce them are also important operational characteristics common to all switch models. Figure 5-3 shows a listing and illustration of the movements and forces for the plunger and lever type switch. A lamp to indicate if the switch has power applied is available on an optional basis.

Limit switches are frequently used to detect the movement of parts or part carriers in an automated work cell. The switch is tripped by a part called a *dog*. The speed at which the part is moving influences the shape

---------- **Definitions of Operating Characteristics** ----------

Operating force (OF):
The force applied to the actuator required to operate the switch contacts.

Releasing force (RF):
The value to which the force on the actuator must be reduced to allow the contacts to return to the normal position.

Free position (FP):
The initial position of the actuator when there is no external force applied.

Operating position (OP):
The position of the actuator at which the contacts snap to the operated contact position.

Releasing position (RP):
The position of the actuator at which the contacts snap from the operated contact position to their normal position.

Total travel position (TTP):
The position of the actuator when it reaches the stopper.

Pretravel (PT):
The distance or angle through which the actuator moves from the free position to the operating position.

Overtravel (OT):
The distance or angle of the actuator movement beyond the operating position.

Movement differential (MD):
The distance or angle from the operating position to the releasing position.

Total travel (TT):
The sum of the pretravel and total overtravel expressed by distance or angle.

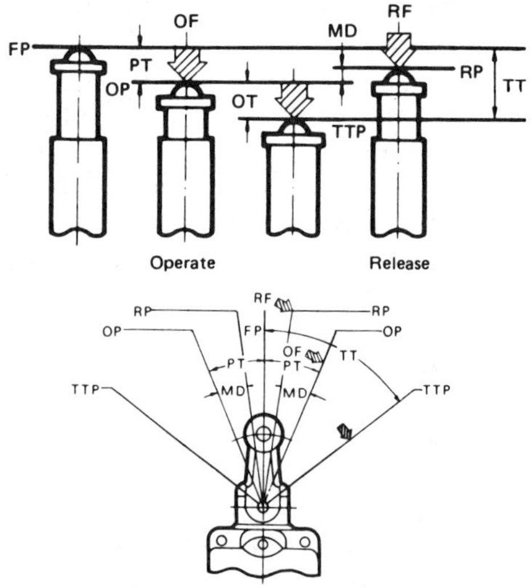

OF:	Operating Force	TTP:	Total Travel Position
RF:	Releasing Force	PT:	Pretravel
FP:	Free Position	OT:	Overtravel
OP:	Operating Position	MD:	Movement Differential
RP:	Releasing Position	TT:	Total Travel

Figure 5-3 Definition of Switch Movement (*Courtesy of Omron Electronics Inc.*)

of the trip *dog* used to actuate the limit switch. The design data given in Figure 5-4 provide data for the sizing of trip *dogs* for lever- and plunger-actuated switches. The non-overtravel *dog* keeps the switch in the actuated position after the switch is engaged. The overtravel *dog* allows the switch to return to the off state after the *dog* has passed. The following steps are used in the design of trip *dogs*:

1. Using the mechanical and electrical specifications of the application, select a receptacle, switch body, and lever or plunger head which meet the requirements.
2. Determine the velocity of the trip *dog* at the time of impact with the lever or plunger of the limit switch.
3. Determine if an overtravel *dog* will be required.
4. Using the data from steps 1, 2, and 3, locate the appropriate velocity table in Figure 5-4.
5. Locate the closest larger table velocity using the value from step 2.
6. Locate the corresponding values of θ, ϕ, and the equation for Y for the table velocity.

- **Design of operating dog**
- **ROLLER LEVER TYPE ACTUATOR**

(1) Non-overtravel dog

(a) V ≤ .82'/sec

If V does not exceed .82'/sec, the lever can be set vertically. Operating speed increases as the dog angle (φ) decreases.

φ	Vmax. (ft/sec)	y
(30°)	(1.31)	0.8 (TT) Dog stroke is allowed up to 80% of TT.
45°	.82	
60°	.33	
(60° ~ 90°)	.16	

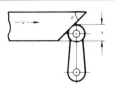

(b) .82'/s ≤ V ≤ 6.56'/s

In considerably high-speed operation, the lever angle (θ) may occasionally be required to change according to the dog angle (φ). It is appropriate to set the lever angle (θ) within a range of 45° to 75° as summarized below.

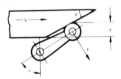

θ	φ	Vmax. (ft/sec)	y
45°	45°	1.64	0
50°	40°	1.97	0.5~0.8 (TT)
60°~(55°)	30°~(35°)	4.26	0.5~0.7 (TT)
75°~(65°)	15°~(25°)	6.56	0.5~0.7 (TT)

NOTE: y is a ratio of the dog stroke to the total travel indicating that the dog stroke is allowed up to 50 to 80% of T.T.

(c) 6.56'/s ≤ V ≤ 9.84'/s

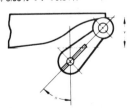

θ	φ	Vmax. (ft/sec)	y
45°	Dog acting-surface is completed with a smooth cubic curve.	9.84	0.5 (T.T)

NOTE: y is a ratio of the dog stroke to the total travel, indicating that the dog stroke is allowed up to 50% of T.T.

(2) Overtravel dog

(a) V ≤ .82'/s

The lever is set vertically. Operating speed increases as the dog angle (φ) decreases.

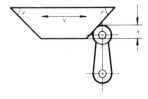

θ	Vmax. (ft/sec)	y
(30°)	(1.31)	0.8 (T.T) Dog stroke is allowed up to 80% of TT.
45°	.82	
60°	.33	
(60° ~ 90°)	.16	

(b) 1.64'/s ≤ V

It is appropriate to set the dog angle (φ) of the rear edge within a range of 15° to 30° or to make the dog acting surface straight to decrease the shaking of the lever.

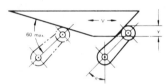

- **PLUNGER TYPE ACTUATOR**

Even with the dog that overrides the actuator, it is allowed that the shapes in the forward direction and backward direction may be the same. However, avoid shapes that will cause the actuator to leave the dog abruptly.

(1) Roller plunger type

φ	Vmax. (ft/sec)	y
30°	.82	(0.6~0.8) T.T
20°	1.64	(0.5~0.7) T.T

NOTE: y is a ratio of the dog stroke to the total travel, indicating that the dog stroke is allowed up to 60 to 80% or 50 to 70% of T.T.

(2) Ball plunger type

φ	Vmax. (ft/sec)	y
30°	.82	(0.6~0.8) T.T
20°	1.64	(0.5~0.7) T.T

(3) Bevel plunger type

φ	Vmax. (ft/sec)	y
30°	.82	(0.6~0.8) T.T
20°	1.64	(0.5~0.7) T.T

Figure 5-4 Trip Dog Design Data
(Courtesy of Omron Electronics Inc.)

7. Determine the Total Travel (TT) for the switch selected as follows:
   For lever type limit switches:

   TT(inches) = Arm Radius(inches) × (1 − Cos(TT in degrees))

   For plunger type limit switches:

   TT(inches) = Value listed in switch data sheet

8. Solve for the trip *dog* stroke value Y using the value of TT from step 7 and the equation from step 6.
9. Draw the trip *dog* and locate the limit switch using the values obtained in steps 8 and 6.

Example Problem:

In a robot welding application a signal is needed to tell the robot when to initiate the welding cycle on the part located on an indexing table. Design a trip *dog* which can be mounted to the edge of the table to actuate a limit switch as the rotary index table is turned into position. The velocity at the outer edge of the table is 0.5 feet per second, and the limit switch selected for the application is model D4A-1201 with a D4A-A00 lever. Non-overtravel type operation is desired in this application.

Solution steps:

1. The switch specified is a D4A-1201 with a D4A-A00 lever. The switch is a standard SPDT model side rotary type with electronic duty, since it will be interfaced to the programmable controller in the robot. No operational indicator lamp is included. Specifications for the switch are given in the Operating Characteristics table under the column marked D4A-☐☐01. The lever specifications are given with the outline drawing under the section labeled levers.
2. The velocity is 0.5 feet per second.
3. Non-overtravel operation is desired for this application.
4. The velocity range for Table 1*a* in Figure 5-4 is 0.16 to 1.31 feet per second, and 0.5 falls within that range.
5. The trip *dog* velocity of 0.5 feet per second falls between 0.33 and 0.82 in the table so the larger value of 0.82 is used.
6. The value of ϕ for this application is 45 degrees and the equation for Y is:

   $Y = 0.8 \times$ Total Travel (inches)

7. TT (Total Travel) = PT (Pretravel) + OT (Overtravel)

   TT = 15 degrees + 60 degrees
   TT = 75 degrees

   TT(inches) = Arm Radius × (1 − Cos(TT in degrees))
   TT = 1.5 inches × (1 − Cos(75))
   TT = 1.5 inches × 0.741
   TT = 1.11 inches

8. $Y = 0.8 \times TT \text{ (inches)}$
    $Y = 0.8 \times 1.11 \text{ inches}$
    $Y = 0.888 \text{ inches}$

9. The trip *dog* and limit switch position to satisfy the problem are drawn in Figure 5-5 using the values obtained.

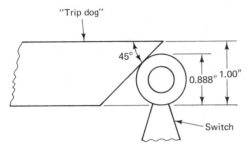

Figure 5-5 Trip Dog for Example Problem

Artificial Skin

Effective tactile sensors and systems are needed for robot grippers used in assembly applications. Research on small and medium size parts-assembly tasks indicates that the development of a basic form of artificial skin would permit robots to perform about 50 percent of the generic tasks found in industrial assembly. Currently most robot interaction with sensors is open loop. For example, gripping devices close on all parts with the same force. A *close gripper* signal is generated by the robot, but feedback of the gripping characteristics to the robot controller is not provided. In grippers with artificial skin, the fingers are closed under servo control with finger pressure and other tactile parameters used for feedback to the controller.

There is an important distinction between *tactile sensing* and *simple touch*. Simple touch includes simple contact or force sensing at one or just a few points on the gripper surface. For example, the gripper in Figure 5-6 uses two switches to determine that the part is centered in the jaws. The switches provide binary data in the form of either an *on* or *off* signal depending on the location of the part. If the switches were replaced with analog pressure pads, the data would be continuously variable and the signal value would become a function of the pressure applied by the fingers. The sensing would still be classified as simple touch, however, since only two contact points were monitored.

Tactile sensing requires a group of sensors arranged in a rectangular or square pattern called an *array*. Figure 5-7 shows gripper fingers with an 8 × 8 tactile-sensing array. The array has 64 sensing elements, each capable of measuring the continuously variable force applied to the element. A tactile sensor interfaced to an intelligent controller can determine the shape,

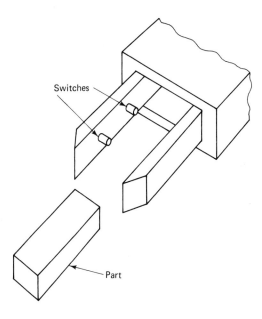

Figure 5-6 Simple Touch Sensing Gripper

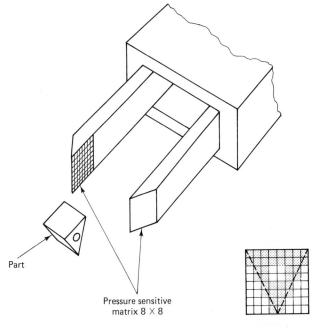

Figure 5-7 Tactile Sensing Array

texture, position, orientation, deformation, center of mass, and presence of torque and slippage of any object held. The array output pictured in Figure 5-7 shows the results of holding a triangular-shaped object in the gripper.

Commercially available tactile sensors can be integrated into assembly robot applications. Figure 5-8 contains a sensor by the Lord Corporation with an 8 × 12 matrix of 96 elements. The load range for each element is 0.05 to 25 pounds for perpendicular forces; shear forces can also be measured. The sensor provides the robot controller with information about the magnitude, direction, and location of the line of action of the total load vector.

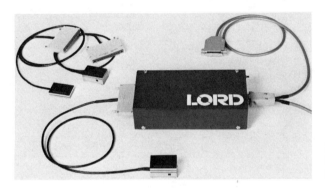

Figure 5-8 Model LTS 200 Tactile Sensor (*Courtesy of Lord Corporation*)

The three steps required for intelligent acquisition of parts by either a robot or a human operator are first *vision*, followed by *proximity sensing*, and finally *tactile sensing*. In order for a robot system to automatically retrieve randomly placed parts or to assemble products at any scale, further development of all three of these sensing areas is required. In addition, if machines are to develop tactile sensing like that found in humans, then a second element must also be incorporated. To the human faculty of skin sensing is added *haptic perception*, which is sensing information that comes from the joints and muscles. Thus robot software must combine the data from the artificial skin with the joint torque information if machines are to achieve what humans call "touch."

5-3 NONCONTACT SENSORS

As the name implies, noncontact sensors measure part characteristics without physically touching the part. In robot work cells noncontact sensors include *proximity* and *photo-optic* devices and *vision* systems. All three approaches have led to the development of commercially available products to solve automated work cell sensing problems.

Classification		HIGH-FREQUENCY OSCILLATION TYPE			
Model		TL-X (Compact type)	TL-X (Plug-in type)	TL-XD	E2P
Features		Cylindrical type proximity sensor boasting ultra small size and high performance	Heavy duty proximity sensor with low operating current and integral receptacle	Cylindrical type proximity sensor with DC-2 wire system	Heavy duty proximity sensor with low operating current and integral receptacle
Appearance & dimensions		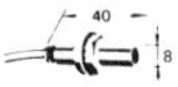 TL-X1□□ 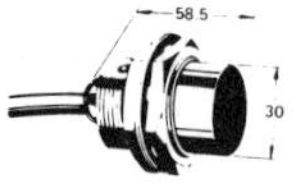 / TL-X18MY□	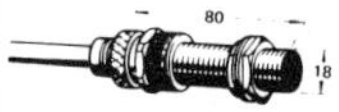 TL-X5Y	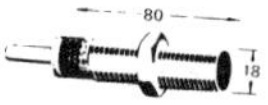 TL-XD(B)5 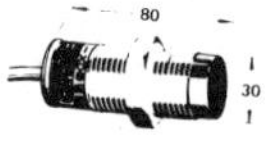 / TL-XD(B)15M	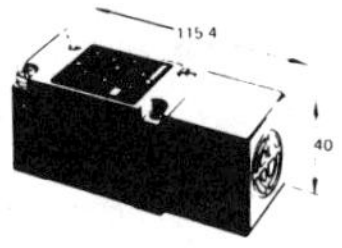
Sensing distance		1, 2, 5, 10, 18mm	5, 10mm	2, 4, 5, 8, 10, 15mm	15mm
Rated voltage	DC switching type	10 to 40 VDC	–	8 to 40 VDC	–
	AC switching type	45 to 260 VAC	90 to 140 VAC	–	90 to 140 VAC
Operating current	DC switching type	10mA max.	–	1.5mA max.	–
	AC switching type	2mA max.	1.7mA max.	–	1.7mA max.
Materials sensed		All metals	All metals	All metals	All metals
Response frequency or time	DC switching type	1kHz min.	–	800Hz min.	–
	AC switching type	20Hz min.	12.5Hz min.	–	12Hz min.
Control output (switching capacity)	DC switching type	200mA max.	–	3 to 200mA	–
	AC switching type	5 to 200mA	5 to 700mA	–	5 to 700mA
Degree of protection	JIS C0920	–	Water-resistant type	–	Water-resistant type
	IEC 144	IP67	IP67	IP67	IP67
	NEMA	Types 1, 4, 6, 12, 13	Types 1, 2, 4, 6, 12, 13	Types 1, 4, 6, 12, 13	Types 1, 2, 4, 6, 12, 13
Ambient temperature		Operating: −40 to +85°C	Operating: −25 to +70°C	Operating: −25 to +70°C	Operating: −25 to +70°C
Approved standards		ⓤ	ⓤ	–	ⓤ
Page		287	293	297	301

Figure 5-9 Proximity Sensors (*Courtesy of Omron Electronics Inc.*)

Proximity Sensors

Proximity sensors detect the presence of a part when the part comes within a specified range of the sensor. Proximity sensors are available in four package shapes with some shapes having several different sizes. The package shapes include *cylindrical, rectangular, through-head type,* and *grooved-head type*. The shape of the part and sensing application dictate the type of package shape for best operation. Figure 5-9 shows the basic sensors available from Omron. The sensors are mounted to supporting brackets using either

Sec. 5-3 Noncontact Sensors

HIGH-FREQUENCY OSCILLATION TYPE	ELECTROSTATIC CAPACITANCE TYPE	HIGH-FREQUENCY OSCILLATION TYPE		
TL-L	E2K	TL-M	TL-N/H/F	TL-YS10
High-performance proximity sensor with 50mm sensing distance & LED for short-circuit indication	Capacitive type senses nonmetalic objects.	Basic switch housing model	High-frequency oscillation type with easy operation monitoring	Direct AC load switching type with easy-to-see operation indicator lamp
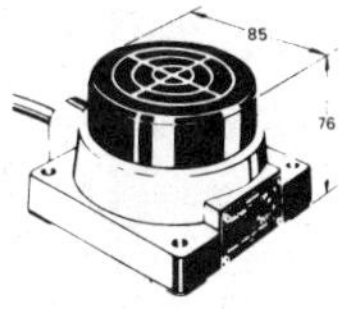 TL-LP50	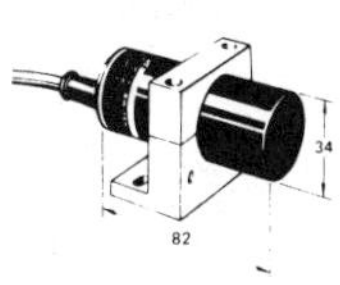		 TL-N5ME□ 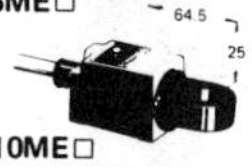 / TL-H10ME□ 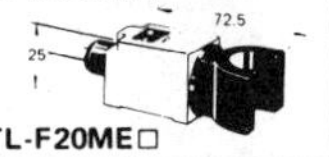 / TL-F20ME□	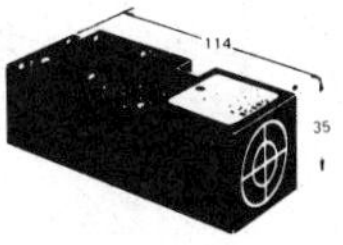
50mm	3 to 25mm	2, 5mm	5, 6, 8, 10, 12, 20mm	10mm
10 to 30 VDC	10 to 40 VDC	10 to 30 VDC	10 to 30 VDC	—
90 to 250 VAC	90 to 250 VAC	90 to 250 VAC	90 to 250 VAC	90 to 130, 180 to 260 VAC
10mA max.	10mA max. (at 12 VDC) 15mA max. (at 24 VDC)	15mA max. (at 24 VDC)	8 mA max. (at 12 VDC) 15mA max. (at 24 VDC)	—
—	1mA max. (at 100 VAC 50/60Hz) 2mA max. (at 200 VAC 50/60Hz)	1.5mA max. (at 100 VAC 50/60Hz) 3.0mA max. (at 200 VAC 50/60Hz)	2mA max.	5mA max.
All metals	Non-metals, Metals	All metals	All metals	All metals
15msec max.	70Hz min.	500Hz min.	500Hz min.	—
25msec max.	10Hz min.	20Hz min.	10Hz min.	10msec max.
200mA max.	200mA max.	100mA max. (at 12 VDC) 200mA max. (at 24 VDC)	200mA max.	—
10 to 200mA	5 to 200mA	10 to 200mA	5 to 200mA	10 to 700mA
	Water-resistant type		—	Water-resistant type
IP67	IP66	IP67	IP67	IP66
Types 1, 4, 6, 12, 13	Types 1, 4, 12, 13	Types 1, 4, 12, 13	Types 1, 4, 6, 12, 13	Types 1, 4, 12, 13
Operating: −25 to +70°C	Operating: −25 to +70°C	Operating: −25 to +70°C	Operating: −25 to +70°C	Operating: −25 to +70°C
ⓤ	ⓤ	—	—	—
305	309	313	317	321

Figure 5-9 (*continued*)

a threaded part of the sensor body or mounting holes. The operating parameters for proximity sensors are defined as follows:

Reference plane: The plane of reference on the proximity sensor from which all measurements are made.

Reference axis: An axis through the sensor from which measurements are made.

Standard object: A definition of the object to be sensed in terms of a specified shape, size, and material composition.

Sensing distance: The distance from the reference plane to the standard object which causes the output of the sensor to change to the *on* state.

Vertical sensing distance: The sensing distance measured by bringing the standard object toward the reference plane with the standard object normal to and centered about the reference axis. (See Figure 5-10.)

Type of sensor	Illustration	Explanation
Column head & square pillar head types	**Vertical sensing distance** (diagram showing reference axis, Releases (ON), Operates (OFF), Reference plane, Object to be sensed, Differential distance, Sensing distance, Proximity sensor)	• The sensing distance is expressed by the distance measured from the reference plane by bringing the object to be sensed closer in the direction of the reference axis (perpendicularly to the sensing surface).

Figure 5-10 Vertical Sensing Distance (*Courtesy of Omron Electronics Inc.*)

Horizontal sensing distance: The sensing distance measured by bringing the standard object along a plane parallel with and at a fixed distance from the reference plane. (See Figure 5-11.)

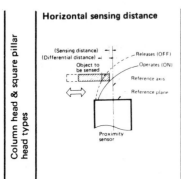

Column head & square pillar head types	**Horizontal sensing distance** (diagram showing Sensing distance, Differential distance, Object to be sensed, Releases (OFF), Operates (ON), Reference axis, Reference plane, Proximity sensor)	• The sensing distance is expressed by the distance measured from the reference axis by moving the object to be sensed parallel to the reference plane (i.e., sensing surface). This distance can be expressed as the locus of operating point, since it varies with the passing position of the object (distance from the reference plane).

Figure 5-11 Horizontal Sensing Distance (*Courtesy of Omron Electronics Inc.*)

Resetting distance: The distance from the reference plane to the standard object which causes the output of the sensor to change from *on* to *off* as the standard object is withdrawn from the sensor. (See Figures 5-10 and 5-11.)

Differential distance: A measure of the hysteresis present in the system. The difference between the resetting distance and the sensing distance for the type of sensor used. (See Figures 5-10 and 5-11.)

Sec. 5-3 Noncontact Sensors   133

Setting distance: The maximum sensing distance when worst-case ambient temperature and supply voltage variations are assumed. (See Figures 5-10 and 5-11.)

Response time: The time required for the output to change states after the standard object's position triggered a change of state.

Frequency response: The maximum rate at which standard objects can cause the output to change states.

Leakage current: The maximum *off* current that will flow from the output terminals with the output stage in the *off* state.

Temperature variation: The variation in sensing distance as a result of variations in ambient temperature.

Voltage variations: The variation in sensing distance as a result of variations in supply voltage.

The selection and the application of proximity devices require an understanding of the *physical* and *electrical* properties of the sensors along with their *operational characteristics*. The reader should refer to the data sheets in Appendix B as these properties and characteristics are described. Note that general reference data are provided along with device data.

All models have solid-state circuits enclosed in the sensing head to generate the fields necessary for remote sensing of objects and production of output signals. The majority of the sensors are designed to detect metal objects, while a few models are available which can detect both metallic and nonmetallic parts. The sensor that detects metal only uses a high-frequency electromagnetic field for detecting part presence, whereas the nonmetallic type uses electrostatic capacitance for the detecting mechanism. Some manufacturers provide three different output circuits for interfacing to other work cell hardware. The output circuits include NPN and PNP in the dc switching type and an ac switching type output. The second page of the Omron data sheet for the TL-X model device in Appendix B shows the three output circuits for that series of sensors. The Omron models also offer the option of normally closed (NC) operation or normally open (NO) action at the output. Most models permit a wide range of voltages for operating power. The model TL-X by Omron, for example, allows the supply power on dc output models to range between 10 and 40 volts dc unregulated, and the ac output devices can use 45 to 260 volts ac. The output can be interfaced directly to a robot controller or connected to a sensor controller provided by the manufacturer. For example, the sensor controller available from Omron is powered from 110 volts ac and provides 12 or 24 volts dc (depending on the model) to power the sensor. The sensor controller also provides relay contact outputs for control of other work cell machines, plus an inverting switch to change between NC or NO operation. The sensor controller, pictured in Figure 5-12, also has two sensor input connections

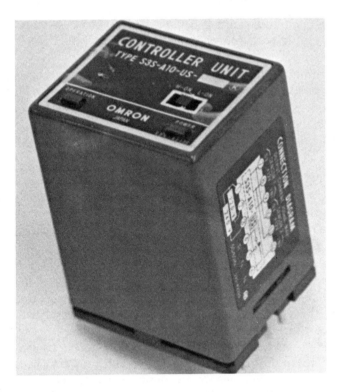

Figure 5-12 Sensor Controller (*Courtesy of Omron Electronics Inc.*)

so that sensor outputs can be logically ANDed or ORed. The Omron data sheet for sensor controllers is included in Appendix B for more detailed study.

The application of proximity sensors is straightforward. *First*, select the sensor model that corresponds to the characteristics of the parts to be sensed. *Second*, mount the sensor so that the object to be detected passes within the sensing distance of the device. *Third*, apply power either through a sensor controller or from an external source. *Fourth*, connect the relay contacts in the sensor controller to a robot or programmable controller which will use the sensor signal for machine control. While this operation is simple, there are numerous design decisions to be made before each of the four steps is taken. In addition, the actual wiring diagram will vary depending on the sensor and controller selected.

Sensing distance is affected by many variables, including the following:

1. type and model of sensor
2. material being sensed
3. path the object uses to trigger the sensor
4. ambient temperature variation

5. supply voltage variation
6. proximity of other objects and other sensors.

For example, an Omron model TL-X10 used to detect a brass plate 50 mm along one side, using a vertical approach, will have a sensing distance of 4 mm. That distance could vary by ± 17.5 percent depending on variations in the temperature and supply voltage. If another proximity sensor were operated next to this unit, it would require a different operating frequency to avoid interference.

Interfacing the power and output signals requires an equal number of design considerations. One factor affecting the sensor selection process is the input characteristics of the machine to which the sensor is connected. A programmable controller or robot controller, for example, will require a low voltage type of sensor output circuit and a motor starter control would require a high power type sensor output. The decision to connect the sensor directly to the machine control or to use a sensor controller will be affected by the availability of power for the sensor and the need for logical operations with two or more sensors. The general reference data from Omron in Appendix B include a set of precautions to follow in selecting a sensor and also some design considerations for the object to be sensed. Review this information carefully.

Photoelectric Sensors

Photoelectric sensors detect the presence of an object or part when the part either breaks a light beam or reflects a beam of light to a receiver. The different types are defined as follows:

> **Separate type:** The sensor system includes two separate devices: a light source to produce a beam of light and a receiving device to sense the presence of the light beam.
>
> **Retroreflective type:** The sensor system includes two separate devices: a sensor with both a light source and receiver present, and a retroreflective target, which is a highly reflective surface. The light beam leaves the sensor, bounces off the target, and then returns to the receiver. The sensor is triggered when a part breaks either the outbound or returning beam.
>
> **Diffuse reflective type:** The sensor has a light source and receiver built in the same case. Stable operation assumes that the part to be detected will return sufficient diffused light to trigger the receiver when the part is in range, and that the background equipment will not return enough light to trigger the receiver when the part is out of range. The light

beam is diffused back to the receiver by using the natural reflective characteristics of the part's surface.

Definite reflective type: This sensor works like a combination of the retroreflective and diffuse reflective types. The light source and receiver are both located in the same enclosure, and the beam uses the part's surface to reflect light back to the receiver. The light reflected must be a definite beam since diffused light alone will not be sufficient to activate the receiver. This sensor is like a retroreflective type which uses the part's surface in place of a target reflector.

The chart in Figure 5-13 shows the four types of photoelectric sensors. The operating parameters used to describe the operation of the sensors are defined as follows:

Sensing distance: For separate and retroreflective types it is the maximum distance between the light source and the receiver (or light source and target) which produces stable operation. For both reflection types it is the maximum distance between the light source and object to be detected that will produce stable operation.

Operating distance: For the reflective type it is the distance from the sensor to the part which causes the output of the sensor to change to the *on* state.

Resetting distance: For the reflective type it is the distance from the sensor to the part which causes the output to change from *on* to *off* as the part is withdrawn from the sensor.

Differential distance: A measure of the hysteresis present in the system. This is the difference between the resetting distance and the operating distance for the type of reflective sensor used.

Optical axis: The axis passing through the sensor along which the light beam is generated or received.

Direction angle: For separate and retroreflective type sensors it is the maximum angle through which the optical axis can move and still provide stable operation.

Dark On operation: For all types of sensors it means that the output will be active or *on* when no light is received by the receiver.

Light On operation: For all types of sensors it means that the output will be active or *on* when light is received by the receiver.

Response time: The time required for the output to change states after the part breaks the beam or reflects the light to the receiver. Response time can be either operate time (time to activate the output) or reset time (time to turn the output *off*).

Sec. 5-3   Noncontact Sensors   137

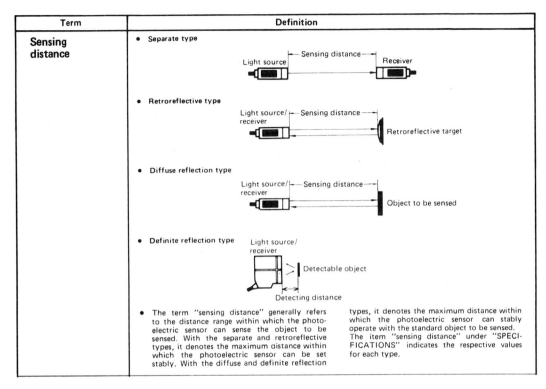

Figure 5-13 Four Types of Photoelectric Sensors (*Courtesy of Omron Electronics Inc.*)

The selection and application of photoelectric sensors require an understanding of the *physical, electrical,* and *optical* properties along with the *operational characteristics* of the different types of sensors. Frequent reference to the data sheets in Appendix B will be useful as these properties and characteristics are discussed. The various types of sensors available from Omron are illustrated in Figure 5-14.

The primary difference between the four types of photoelectric sensors is in their range, that is, distance to the part detected. The separate type sensor system permits the light source and receiver to be separated by up to 30 meters for some models, so that detection can be anywhere along the 30-meter length. The retroreflective type has separation distances between the sensor and target from 1 to 5 meters. The diffused-light type has a detection distance which is less than 1 meter and typically in the 5- to 50-centimeter range. Last, the definite reflective type has the shortest detection distance—typically from 5 to 25 centimeters. While the range is small for the definite type, its ability to operate with high light reflection from back-

Classification	AMPLIFIER SELF-CONTAINED TYPE			
Model	E3A	E3B	E3N	E3S
Features	Slim styled photoelectric sensor with built-in amplifier	Slim styled photoelectric sensor with built-in relay output in rigid die-cast case	Built-in amplifer type with convenient bicolor indicator	Miniature water-resistant photoelectric sensor with built-in amplifier
Appearance & dimensions	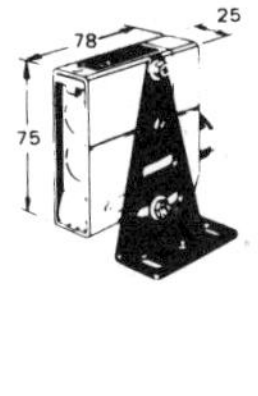	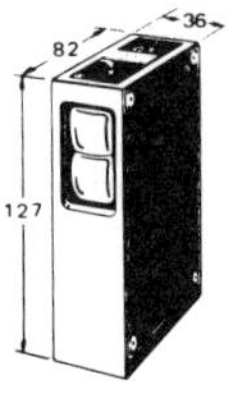	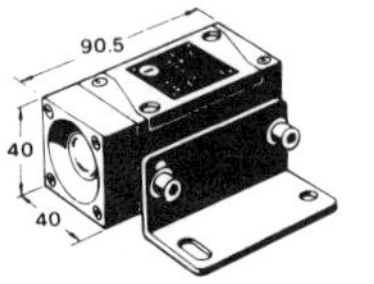	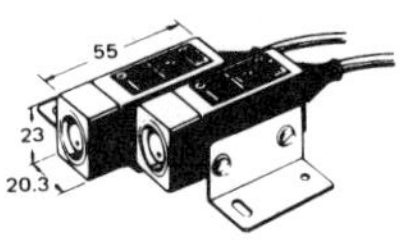
Method of sensing	Separate type Retroreflective type Diffuse reflection type	Separate type Retroreflective type Diffuse reflection type	Separate type Retroreflective type Diffuse reflection type	Separate type Retroreflective type Diffuse reflection type
Rated voltage	90 to 250 VAC, 50/60Hz	120 VAC, 50/60Hz	12 VDC −10% to 24 VDC +10%	12 VDC −10% to 24 VDC +10%
Operating current	2, 3, 4.5VA max.	4VA max.	60 or 80mA max.	40mA or 50mA max.
Sensing distance	0.1 to 5m	1, 5 or 10m	50cm, 2, 5, 10 or 30m	1.2 ±0.2, 5 or 20cm 0.1 to 1, 1 or 3m
Object to be sensed	Transparent and opaque materials	Transparent and opaque materials	Transparent and opaque materials	Transparent and opaque materials
Control output	Contact output (SPDT) 250 VAC 1A Solid-state output (SCR) Switching capacity: 250 VAC, 200mA max.	Pilot duty 120/240 VAC 180VA, 5A 120/240 VAC resistive	Solid-state output: Output (source) current: 1.5 to 3mA Load (sink) current: 200mA max.	Solid-state output: Output (source) current: 1.5 to 3mA Load (sink) current: 80mA max.
Response time	Comtact output: 60msec max. SCR output: 30msec max.	30msec max.	5msec max.	5msec max.
Ambient temperature	Operating: −10 to +55°C	Operating: −10 to +55°C	Operating: −25 to +55°C	Operating: −25 to +55°C
Degree of protection — JIS (C 0920)	Water-resistant type	Water-resistant type	Drip-proof type	Water-resistant type
Degree of protection — IEC144	IP66	IP66	IP65	IP62 or IP66
Degree of protection — NEMA	Types 1, 4, 4X, 12	Types 1, 4, 4X, 13	Types 1, 2, 12	Types 1, 2 or 1, 4, 4X, 12
Approved standards	—	(UL) (CSA)	(UL) *	(UL) *
Page	331	335	339	345

NOTE: * DC switching type only.

Figure 5-14 Photoelectric Sensors (*Courtesy of Omron Electronics Inc.*)

ground objects is excellent and far superior to the diffused reflective type under those conditions.

The sensor unit consists of one or two packages, depending on the model and manufacturer, with mounting brackets designed for easy alignment of the sensors. Figure 5-15 shows the internal arrangement of parts and components of an Omron model E3B sensor. The packages are designed for operation under varying environmental conditions, and models can be found which meet most NEMA standards. The sensors include solid-state sensing electronics, which drive the light source in the transmitter, and a light-sensitive element in the receiver. The electronics require dc or ac voltage for power (typically 12 to 24 volts regulated dc or 90 to 250 volts ac). The sensor output includes an electrical signal and light-emitting diode to

Sec. 5-3 Noncontact Sensors   139

AMPLIFIER SELF-CONTAINED TYPE			AMPLIFIER SEPARATED TYPE	Color mark sensor
E3S-G	E3S-L	E3S-X	OPE/ORE	E3ML
Grooved head type ideal for mark sensing and positioning	Focusable type with built-in amplifier	Fiber optics type for high resolution sensing	Subminiature series ideal for OEM use	High-performance color mark sensor
Grooved head type	Definite reflection type	Separate type Diffuse reflection type	Separate type Diffuse reflection type	Separate type Diffuse reflection type
12 VDC −10% to 24 VDC +10%	12 VDC −10% to 24 VDC +10%	12 VDC −10% to 24 VDC +10%	120/240 VAC ±10%, 50/60Hz (with amplifier unit)	10 to 30 VDC (For lamp 4.5 VAC)
40mA max.	50mA max.	50mA max.	Approx. 5VA	40mA max. (at 30 VDC)
5mm, 3cm	3 to 10cm or 5 to 25cm	8mm, 3cm	3, 5 or 10cm 1 or 3m	0.5 to 20mm
Marks on transparent sheet Opaque and translucent materials	Transparent and opaque materials	Transparent, translucent and opaque materials	Transparent, translucent and opaque materials	Any color mark
Solid-state output: Output (source) current: 1.5 to 3mA Load (sink) current: 80mA max.	Solid-state output: Output (source) current: 1.5 to 3mA Load (sink) current: 80mA max.	Solid-state output: Output (source) current: 1.5 to 3mA Load (sink) current: 80mA max.	Contact output: 230 VAC 1A Solid-state output: 15 VDC Output impedance: 3.3kΩ	Solid-state output: Load current: 80mA max. Output impedance: 4.7kΩ
2msec max.	5msec max.	2msec max.	Contact output: 30msec max. Solid-state output: 3msec max.	20μsec max.
Operating: −25 to +55°C	Operating: −25 to +55°C	Operating: −25 to +55°C	Operating: −25 to +55°C	Operating: −10 to 55°C
Water-resistant type	Water-resistant type	Water-resistant type	Dust-proof type	Immersion-proof type
IP66	IP66	IP66	IP51	IP67
Types 1, 4, 4X, 12	Types 1, 4, 4X, 12	Types 1, 4, 4X, 12	—	Types 1, 4, 4X, 12
ⓊⓁ —	ⓊⓁ *	ⓊⓁ —	ⓊⓁ *	—
351	355	361	365	371

Figure 5-14 (*continued*)

indicate the condition of the sensor. The electrical signal can drive a control amplifier which provides relay contacts for use in switching higher current loads or can drive smaller loads directly. The Omron models, for example, have direct output drive capability which ranges from 80 milliamperes on some models to 200 milliamperes on the larger units. The response time of photoelectric sensors is in the 2- to 30-millisecond range.

The sensitivity of photoelectric sensors can be adjusted so that sensing of transparent, translucent, and opaque materials is possible with the same model sensor. The procedure for adjustment of sensor sensitivity varies according to models and manufacturers. Several examples of setup procedures recommended by Omron are included in Appendix B for your review.

Figure 5-16 shows several examples of sensors' being used to check

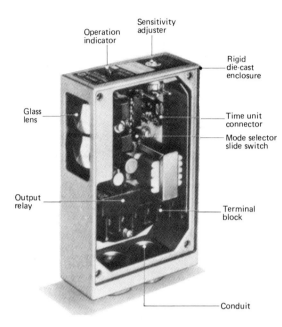

Figure 5-15 Sensor Construction (*Courtesy of Omron Electronics Inc.*)

for the presence of parts. These could very easily be incorporated into a robot work cell. For example, Figure 5-16a,c,g,j, and k illustrate how a sensor could be used to check for the presence of various types of parts a robot might be moving in a material handling application or using in an assembly application. The application in b involves the task of unstacking thin plates. The signal from the sensor is used by the robot controller to stop the downward motion of the arm and activate a vacuum gripper. In a robot drilling application, the drill length must be checked frequently to be sure a drill rod has not been broken. The application in d shows how this is accomplished with a photoelectric sensor.

Application of photoelectric sensors is straightforward. *First*, select the model sensor which satisfies the requirements of the work cell. *Second*, mount the sensor in the work cell and adjust the sensitivity setting with a part in position. *Third*, apply power to the sensor either from a sensor controller or from any external source of dc regulated power. *Last*, interface the sensor to the robot controller or a programmable controller either directly or through a sensor controller which provides relay contacts.

The factors affecting sensor selection include:

1. sensing distance required by the application
2. sensor mounting requirement
3. work cell area available
4. the size, shape, and surface reflectivity of the part to be sensed
5. response time required

Sec. 5-3   Noncontact Sensors

- When the sensor is susceptible to the reflection from background object surface

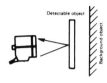

Typical examples
(1) Sensing of thin objects on the conveyor line.
(2) Sensing of objects in the presence of a background object with high reflection factor such as rollers, metallic plates, etc.
(3) Sensing of the residual quantity in a hopper or a parts feeder.

(A)

- Sensing of level or height

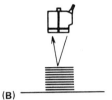

Typical examples
(1) Sensing the height of stacked plywood, tiles, etc. from above.
(2) Monitoring and control of the liquid level from above.
(3) Determination of the heights of objects on a conveyor line.
(4) Sensing of slack in sheets from above.

(B)

- Sensing of objects traveling in contiguous succession

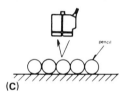

Typical examples
(1) One-by-one sensing of pencils or metallic bars traveling successively or in contiguous succession.
(2) Similarly, one-by-one lateral sensing of bottles or cans traveling in contiguous succession.

(C)

- Sensing of small, slender or fine objects

Typical examples
(1) Sensing of broken drill bits.
(2) Sensing of small parts such as electronic components.
(3) Sensing of the presence or absence of bottlecaps.
(4) Sensing of fine mesh.

(D)

- Sensing of small holes, narrow openings, or unevenness

Typical examples
(1) Sensing of holes in flat board.
(2) Sensing of protrusions.

(E)

- Sensing of objects utilizing their difference in luster

Typical examples
(1) Identifying the face or back of tiles.
(2) Identifying the face or back of lids.

(F)

- Sensing of transparent objects

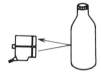

Typical examples
(1) Sensing of transparent or translucent objects.
(2) Sensing of transparent glasses, film or plastic plates.
(3) Sensing of the liquid level.

(G)

- Sensing of objects through a transparent cover

Typical examples
(1) Sensing of the contents in a transparent case.
(2) Sensing of the position of meter pointer.

(H)

- Sensing of the edge of object

Typical examples
(1) Positioning control of plywood
(2) Positioning control of various other products.

(I)

- For sensing of presence of parts in parts feeder

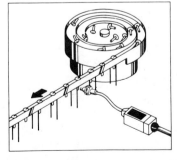

(J)

- For sensing of presence of resistors on conveyor line

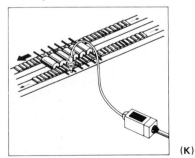

(K)

Figure 5-16 Example Applications of Photoelectric Sensors (*Courtesy of Omron Electronics Inc.*)

6. environmental conditions in the work cell, especially background light present
7. the interface requirements.

Selecting the sensor which best satisfies the work cell conditions requires a thorough study of the application. The primary elements of the study are the factors listed above, and the results of the study are criteria which can be used to select the sensor model for the job.

By means of a sensitivity adjustment on the sensor, the same model device can detect parts with different surface reflectivity from the same sensing distance. The sensitivity is adjusted by following a setup procedure on the sensor data sheets. On some models the operation indicator light-emitting diode (LED) changes colors as the sensitivity is adjusted; on others, the indicator LED changes from *on* to *off* to indicate the sensing condition. On all models with a sensitivity adjustment, the sensor can be operated in a stable mode with the proper setting.

Interfacing the sensor to supply power and controller inputs requires the designer first to consider the model sensor selected. If regulated dc between 12 and 24 volts is available, then a sensor controller is not required. However, without a sensor controller the sensor output must be interfaced directly to the robot controller input or the input of a programmable controller (PC). This requires an output circuit in the sensor which is compatible to the input circuit in the robot or PC.

Vision Systems

The last type of noncontact sensor commonly used in robot work cells is actually a complete intelligence system designed to give the work cell eyes. About 700 vision systems are currently used in automated work cells to perform the following tasks:

Part identification: Commercially available vision systems store data for different parts in active memory and then use these data to distinguish between parts as they enter the work cell. One system, the VS-110 from Machine Intelligence Corporation, learns the characteristics of nine different parts and identifies each part from its two-dimensional silhouette. At the same time, the system rejects all other parts which are not exactly like the nine it was taught.

Part location: Current vision technology allows the user to locate randomly placed parts on an X-Y grid. The VS-110 system, for example, measures the X and Y distances from the center of the camera coordinate system to the center of the randomly placed part. Such systems measure and calculate numerous parameter values which aid in the location of parts presented to a work cell in a random fashion.

Part orientation: Every part must be gripped ¡
the end-of-arm tooling. The vision system s
formation and data which are used to driv⌐
orientation for part pickup. Numerous pai.
both measured and calculated, are provided by th.
in automated parts handling.

Part inspection: Vision systems are used to check parts i.
accuracy and geometrical integrity. The parts are measu.
camera and the dimensions are calculated; at the same time, th.
system checks the parts for any missing holes or changes in the ₊
geometry.

Range finding: In a few applications the system uses two or more cameras to measure the distance from the cameras to the part. This technique is used to measure and calculate the cross-sectional area of parts.

The use of vision to enhance the operation of an automated work cell has moved from the research laboratory to the factory floor. Over 15 manufacturers provide equipment that gives robots the eyes they need to perform complex manufacturing tasks. It is estimated that 50,000 vision systems will be installed in work cells by 1992. The majority of these will be used to give robots the part identification, location, and orientation information necessary for automatic handling of randomly delivered parts.

Vision System Components

The block diagram in Figure 5-17 shows the typical vision system components, which include

1. one or more cameras
2. a camera controller
3. interface circuits for the camera and work cell equipment
4. a high-resolution cathode ray tube (CRT) for image display.

The cameras use a lens system and a charge coupled device (CCD) array or vidicon tube to measure the level of light reaching the camera from the silhouetted part under study. The acquisition of the images is the responsibility of the vision camera and camera controller. Frequently this part of the system is produced by a different manufacturer from the one supplying the vision-processing hardware and software. The vision cameras currently available use two different techniques to capture an image.

The first is a CCD camera which uses a solid-state array of light-sensitive cells deposited on an integrated circuit substrate. Each cell is a small light-

Sensors and Interfacing   Chap. 5

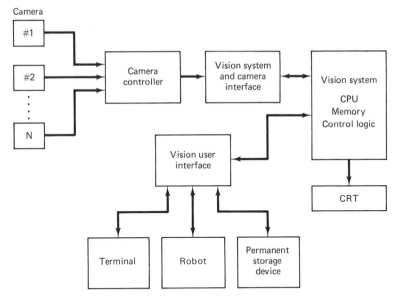

Figure 5-17 Vision System Block Diagram

sensitive device whose output is a function of the intensity of light striking its surface. The CCD systems come in two basic configurations: *linear* arrays and *imaging* arrays. Figure 5-18 shows how each of the types is organized. The linear arrays measure a single line while the imaging arrays measure a complete two-dimensional image. CCDs are very accurate, rugged, and linear but are the more expensive choice when a large number of elements is required in the array.

The second type of camera uses a vidicon tube to measure the light level from the part. For large arrays the tubes are more economical, but they are not as linear or stable and can have a ten percent error due to distortion.

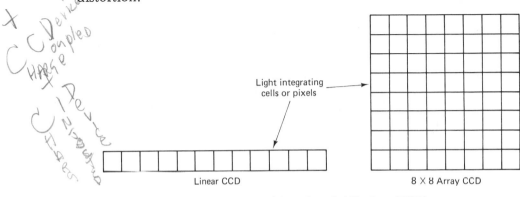

Figure 5-18 Charge Coupled Devices (CCD)

Sec. 5-3   Noncontact Sensors

Image Measurement

The basic unit of measurement in a vision system is the *grey scale*; the basic parameter measured is *light intensity*; and the basic measurement element is the *pixel*. The vision system lens focuses light from the part onto the light-sensitive surface in the camera. The light-sensitive surface is divided up into small regions or picture cells. Each of these regions or picture cells is called a pixel. The resolution of the vision system is directly proportional to the number of pixels on the light-sensitive surface. For example, a sensor with a 256 × 256 array will have a higher resolution than a sensor with a 128 × 128 array of pixels. Higher resolution means greater accuracy when a system is used to measure part dimensions.

Each pixel on the sensor's surface is excited by the light focused on it by the camera lens. With no light present the pixel is turned *off*, but when light reaches a *saturation level* the pixel is *on*. Between those two extremes there are shades of grey which cause the pixel to be excited to a partially *on* condition. The number of excitation states between *off* and *on* is called the grey scale. Current systems have grey scales ranging from 4 to 64. The greater the number in the grey scale, the better the system is at locating fuzzy edges and shadows of parts. The grey scale of a pixel is the numerical representation of brightness for one small spot on the part under the lens.

Image Analysis

The first step in image analysis is to locate all the areas in the image which correspond to the part being viewed by the camera. Most current vision systems are two-dimensional, which means that the outline of the object is the only feature of interest to the system. The two-dimensional features of a part are amplified by placing the part on a back-lighted surface or by placing the object on a surface with contrasting color and using top lighting. Figure 5-19a shows a part on a back-lighted surface, and b shows the corresponding image generated by the vision system. The two most commonly used systems to find objects from camera data are *edge detection* and *clustering*. Both of the techniques attempt to locate the boundaries of objects or regions in the image so that their location, size, shape, and orientation can be computed as clues for recognition. Edge detection is based on the fact that there is a sharp difference in brightness between the object and its background. This can be seen in Figure 5-19a. Areas of high contrast are found by searching through the array of pixel data for jumps in the grey level. To locate the change in grey level a mathematical operator is used which examines a small neighborhood of adjacent pixels and computes an *edge probability* value. If the value is above a set threshold, then a notation is stored in the data indicating a probable part edge. The ability of the system to eliminate false edges due to electrical noise and to find fuzzy edges

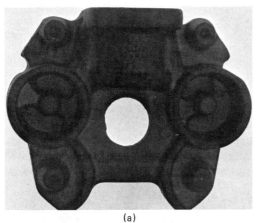 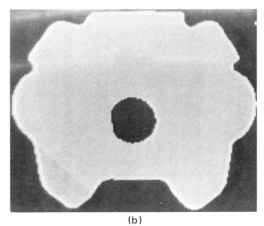

(a)   (b)

Figure 5-19 Vision System Representation of a Production Casting

is improved if the size of the neighborhood used to compute the *edge probability* has a large number of pixels. Since increasing the size of the neighborhood increases the computation time for the system, a trade-off is required.

The second method used to find objects from an array of pixel data is called *clustering* or *region growing*. The system basically tries to find adjacent pixels that have similar properties. This technique uses a *discrimination function* to determine if a pixel is part of a region being developed. The discrimination function computes the desired properties of a pixel and compares them with the properties of other pixels in the region or against a threshold value. In the simplest case the grey level of each pixel could be checked against a threshold value to determine if the pixel is in the part image or in the background area. Many vision systems locate parts by using every pixel whose grey level value is different from the grey level value for the background. This process starts with a *seed* pixel within the part image and then grows in every direction until a boundary is reached. After the area of the part has been developed, the pixel data are frequently changed to binary values; that is, every pixel which is part of the object is stored as a *one* and every pixel which is not part of the object is stored as a *zero*. Dumping these data to a CRT produces the part silhouette shown in Figure 5-19b.

This technique uses more memory locations than the *chain code* or *compressed line* method. The *chain code* technique is illustrated in Figure 5-20. Note that the image edge is formed by four different short vectors which are identified by four different numbers. The object's edge is stored as a series of numbers or codes representing the vectors required to form the outer boundary. The *compressed line* technique passes a horizontal line through the image and finds the points where the line intersects the object's boundary. These points are stored for each horizontal line passed through

Sec. 5-3   Noncontact Sensors   147

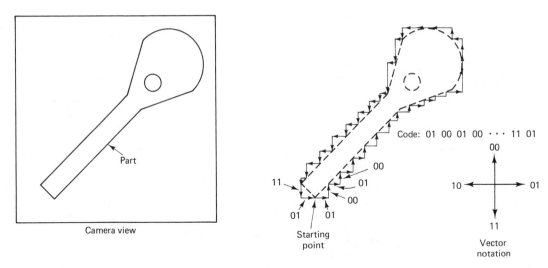

Figure 5-20 Chain Code Technique

the image. These last two techniques require additional processing time but reduce the memory required to store the image sensed.

Image Recognition

The second problem facing the vision system after the edges are detected and stored is the *recognition* or *identification* of the current image. Three frequently used two-dimensional recognition strategies are (1) template matching, (2) edge and region statistics, and (3) statistical matching with the SRI algorithm.

In the *template matching* technique templates of the parts to be recognized are stored in the vision system memory. The images recorded by the vision camera are compared to the templates stored in memory to determine if a match part is present. Figure 5-21 shows the template of a part and the

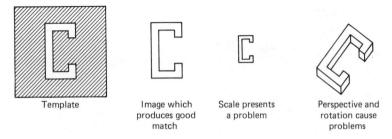

Figure 5-21 Template Matching

problem which this type of system must overcome. Any change in the object's scale or a rotation of the view makes a match with the template more difficult.

The *edge and region statistics* technique defines significant features for the parts studied and then develops a method of evaluating these features from the pixel data. Some of the more common features used in this method are:

> **Center of area:** A unique point from which all other points on the object are referenced.
> **Major axis:** The major axis of an equivalent or best fit ellipse.
> **Minor axis:** The minor axis of an equivalent or best fit ellipse.
> **Number of holes:** The number of holes in the object's interior.
> **Angular relationships:** The angular relationship of all major features to one another.
> **Perimeter squared divided by the area:** A value unique to any given shape that does not change as the image of the object changes scale.
> **Surface texture:** An identifying characteristic where grey level images are used.

The statistical data for parts to be identified are stored in memory, and when the camera produces an image of an unknown part the system calculates a set of feature values from the pixel data. The feature values for the unknown part are compared statistically to the feature values stored in memory. If a close match is found, then the part is recognized and identified by the system.

The *Stanford Research Institute* (SRI) technique is very similar to the method just described. The one exception is that an algorithm is used to apply the features identified for each part. The algorithm includes the following steps:

1. **Setup mode:** The camera is selected, the pixel thresholds are set, and the system is calibrated.
2. **Training mode:** The system is taught the parts which are to be identified by repeatedly viewing each part and calculating the average feature values. All features are calculated even if some will not be used later for identification.
3. **Tuning system for object set:** The part features which would discriminate best between the parts are chosen for use during system operation. The threshold for matching confidence is selected.

4. **Operation:** Parts are viewed by the camera and the features calculated from pixel data. The selected features are compared to the features of the taught points and an identification judgment is made.

Vision will be an important tool in future automated work cells, but a significant amount of work remains to be done before it can perform the basic operations now done by human operators. The areas requiring additional work include three-dimensional camera systems, ranging techniques, selecting parts from randomly stacked parts bins (called the "bin-picking problem"), and handling missing or extra features in taught objects.

5-4 PROCESS SENSORS

In addition to the sensors required for the robot, most process or manufacturing operations need sensors to monitor parameters inherent in the process itself. Also, many of the other automated machines have sensors to alert and warn the operator about conditions which are developing within the operation. A forging operation is a good example. Through warning lights the human operator monitors the temperature of the oven that is heating the parts before forging. The level of oil and oil pressure for the press are also displayed by indicators. With a robot present, the visual indicators are ignored, and important process or machine warnings may not be used to correct out-of-tolerance conditions. These same signals can instead be used to signal the robot or work cell controller that a situation exists that needs corrective action.

The only requirement for the system or machine parameter to be monitored is that it must be discrete, that is, either an *on* or *off* signal. Analog sensors which have a variable output, zero through five volts, for example, are associated with proportional control requirements. While these could be interfaced to the work cell or robot controller, they are usually handled by a separate control system designed to provide proportional control. A discrete signal could be provided by the proportional controller to warn the robot that the process is off-course and a correction is necessary.

In most automated work cells the robot controller acts as the overall system monitor. All sensors and discrete signals are interfaced to the controller input/output (I/O), and the program is designed to include these *go/no-go* signals in the control of the work cell. Figure 5-22 shows the layout of a forging operation and includes a listing of the parameters sensed and the types of sensors used. The signals which are provided by machines used in the work cell are indicated.

150 Sensors and Interfacing Chap. 5

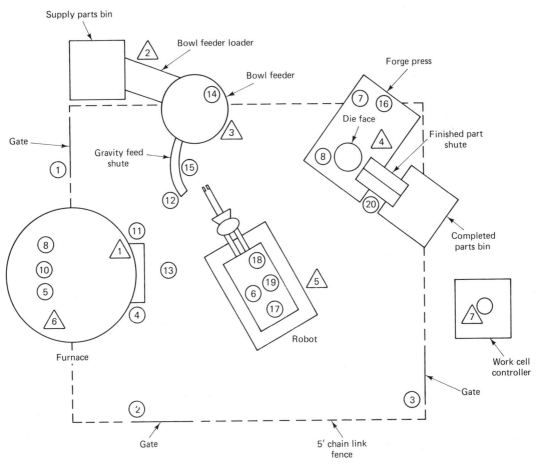

Note:

△ — Signals from the controller for work cell equipment

○ — Work cell sensors

SENSORS TO PROTECT HUMANS FROM HARM

Sensor Number	Parameter Sensed	Sensor Type	Operation Notes
1	Gate open	Limit switch	When gate open switch is cycled, the system goes to a halt condition and must be restarted from outside the workcell with the gate closed.
2	Gate open	Limit switch	
3	Gate open	Limit switch	

Figure 5-22 Example Robot Work Cell

Sec. 5-5   Interfaces

SENSORS TO PROTECT THE ROBOT AND OTHER EQUIPMENT FROM HARM

4	Furnace door closed	Limit switch	
5	Parts holder not indexed	Limit switch	Furnace holds up to 20 parts on rotating rack.
6	Arm in retracted position	Proximity	Sensor determines if the arm is clear of press and furnace.
7	Press open	Limit switch	Press is ready for loading.
8	Part not ejected	Photo electric	Die is not ready for new part.

SENSORS TO MONITOR PRODUCTION SYSTEM FOR PRODUCT QUALITY

9	Furnace temperature too high	Temperature switch
10	Furnace temperature too low	Temperature switch
11	Furnace door open	Limit switch
12	Part not oriented for pick-up	Photo electric

SENSORS TO MONITOR PRODUCTION SYSTEM FOR EQUIPMENT MALFUNCTION

13	Part not in gripper	Photo electric
14	Bowl feeder low on parts	Proximity
15	Feed shute low on parts	Photo electric
16	Press cycle incomplete	Limit switch
17	Robot hydraulic oil low	Pressure switch
18	Robot hydraulic oil too hot	Temperature switch
19	Robot cycle stopped	Limit switch

SENSORS TO ANALYZE PRODUCT QUALITY

20	Part counter	Photo electric	One part out of 100 is inspected. The counter keeps a count so an inspector can be alerted.

SIGNALS PROVIDED BY THE CONTROLLER FOR WORK CELL EQUIPMENT

1	Open furnace door
2	Load bowl feeder
3	Start bowl feeder
4	Start press cycle
5	Start robot cycle
6	Index furnace parts holder
7	Turn on alarm light and horn

Figure 5-22 (cont.)

5-5 INTERFACES

The very nature of a robot system dictates that it must work with other equipment or parts in a manufacturing system. If this is so, then it must be interfaced to the other hardware.

An interface *is defined as a place at which independent systems meet and act on or communicate with each other.*

In a robot system there are several different independent systems which must be interfaced so that communication between systems is possible. The numerous independent systems are grouped into four interface categories called *simple sensor interface, wrist interface, common robot control interface*, and *complex sensor interface*. All interface requirements of the robot will fall into one of these categories.

Simple Sensor Interface

Currently the most well-defined interface area is the simple sensor interface. A *simple sensor* is basically one which has its signal originate in some peripheral hardware or device. Peripheral hardware is defined as equipment used in conjunction with an industrial robot in the design of a work cell. An additional requirement stipulates that the communication with the robot must be by discrete signals only (*on* or *off*). The group does not include peripheral devices such as disk drives, printers, or equipment using higher frequency binary-coded signals with a data format. Equipment in this group does include all the discrete sensors, such as limit switches, proximity sensors, and photoelectric sensors. In addition, all discrete process signals to and from the controller are included. These would be signals from machine tools, welders, and material handlers, plus the signals that the controller originates to operate these peripherals.

The standard logic signal levels for this interface are zero volts for the low and 110 volts ac or 24 volts dc for the high level. Most controllers pro-

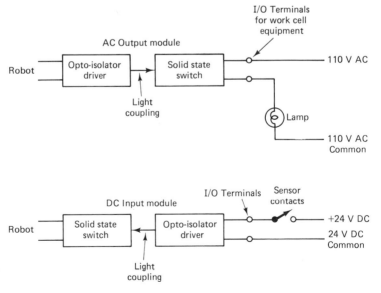

Figure 5-23 Simple Sensor Interface in Robot Controller

Sec. 5-5 Interfaces 153

vide input/output (I/O) modules compatible with these levels. The I/O uses optical coupling between the robot and the peripheral to assure isolation of power and grounds between the different systems. Figure 5-23 illustrates a typical input and output module for a robot with the typical wiring required. Figure 5-24 shows the I/O interface in the Cincinnati Milacron controller used on all the electric models. Note in Figure 5-23 that the only connection between the robot and the external circuit is the light which activates the solid-state switch. The output module acts as a switch to apply power to the external load. The input module has one lead connected to

Figure 5-24 I/O Modules in the Robot Controller

the electrical common and the other lead connected to the external sensor which switches power to the input and activates the light coupling and the solid-state switch. The opto-isolated driver is usually a light-emitting diode, and the solid-state switch is a light-activated transistor for dc or a triac for ac modules.

Electrical noise from industrial machines and contact bounce on the sensor switch contacts can cause I/O problems in the system. Adopting the following guidelines will eliminate many of the potential problems.

1. Eliminate ground loops and provide good grounds.
2. Shield wires when excessive electrical noise is present.
3. Use damping in the sensor contacts or delays in the robot program to avoid the multiple signals produced by bouncing contacts.
4. Use only opto-isolated I/O modules.
5. Use arc suppression circuits to reduce noise from switching inductive devices.
6. Route 110 volt ac and 24 volt dc signal lines in separate wire bundles.

No standard connector type or size is recommended for the simple sensor interface. Most manufacturers use screw-type terminal strips for termination of signal wires.

Wrist Interface

Past applications of robots have been in dedicated, high-volume tasks where a single gripper would satisfy production requirements. The end-of-arm tooling is usually a unique design specifically built for the chosen task. In these applications the interface between the wrist and tooling includes bolt-type fasteners for mechanical linkage and permanent connections for the electrical and pneumatic/hydraulic linkages. With this type of interface between the robot tool plate and the end-of-arm tooling, automatic or even rapid changing of the tooling is not possible. As the need for flexible automation grows so will the requirement for a flexible interface for the end-of-arm tooling. The *wrist interface* for flexible manufacturing must satisfy the following interface requirements:

Mechanical interface: The robot must be able to change tooling under program control, and the integrity of the mechanical linkage must be as good as that experienced with threaded fasteners. The interface must provide both registration and orientation control from one tool to the next.
Electrical interface: The electrical signals used for control of the tooling or signals coming from sensors mounted on the gripper must be

separated automatically when the tooling is changed. For example, the power to electrically powered tools must be disconnected and then reconnected as the tooling is changed.

Hydraulic and pneumatic interface: The same rationale for a quick-breaking interface that was developed in the electrical area will also apply here.

Replaceable or quick-change capability: Future applications must provide for rapid tool change by the user and the robot.

The advantages of a standard wrist interface include:

- The same tooling can be used on every robot in a manufacturing facility without special adapters for each robot.
- No program editing or reprogramming is required when a worn or damaged gripper is replaced.
- Off-the-shelf tooling for general applications could be developed and sold.

No standards are currently available to guide the design of a wrist interface, and very few examples of complete wrist interfaces exist in industrial applications. One design developed by McDonnell Douglas Corporation and funded under the Air Force ICAM project had a complete mechanical, electrical, and pneumatic interface.

Robot Control Interface

This is a robot-independent control interface that allows real-time trajectory control of the robot arm. This interface gives the user the option to control the robot with programs generated external to the robot manufacturer's controller. The interface permits control of any robot model within a specific classification in a real-time mode. The real-time operation permits a computer, external to the robot controller, to change arm motion and programmed points while the program is running. There are three entry points where this interface could connect to the robot controller. These entry points, illustrated in Figure 5-25, are the *trajectory control* level, *coordinate transform* level, and *joint control* level.

Joint Control Level

The lowest level of control in this interface uses positional information in joint space coordinates. This type of control system, the simplest available on servo-driven robots, uses stored joint angles to position the robot arm. The robot is programmed by moving the arm joints until the tooling is in

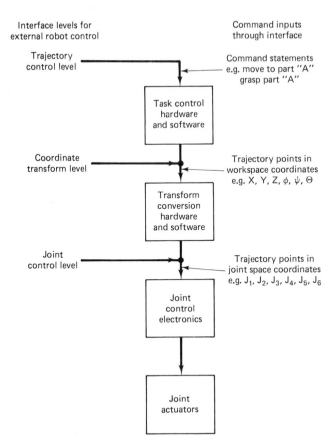

Figure 5-25 Interface Levels of External Robot Control

the correct position. When the program button is pressed, the current joint angles are recorded in memory. A sequence of recorded points or joint angles forms a complete program. The program is executed when the controller uses the sequence of stored points to drive the robot through the required joint angles. The robot can perform a variety of prerecorded tasks by replaying different sequences of these joint space coordinates. The major disadvantage of this level of control is that programs are not transferable between robots. The joint angle data are repeatable for only a single arm, since a specific arm was used to generate the joint space coordinates in the teaching mode.

Coordinate Transform Level

Controllers with coordinate-transformation capability offer the user characteristics such as straight-line motion, tool-center-point control, and other motion control which references a robot-independent coordinate system.

Sec. 5-5   Interfaces

In addition, programming is easier since the tool-center-point is controlled while the robot is moved in the *rectangular*, *cylindrical*, or *hand* coordinate systems. Simpler programming is also a result of joy stick control of the tool movement on some models. With a joy stick, the programmer's only concern is the direction the tool-center-point must move to get to the correct position.

The heart of this control system is a coordinate-transformation algorithm which defines the relationship between the coordinates of the work cell in Cartesian X, Y, and Z values and the joint angles of the arm, represented by $J1$, $J2$, $J3$, $J4$, $J5$, and $J6$, in the joint space. With the additional computational power and the transformation algorithm, the user works in Cartesian coordinates and the controller calculates the necessary joint angles to achieve the desired results. Figure 5-26 illustrates this process.

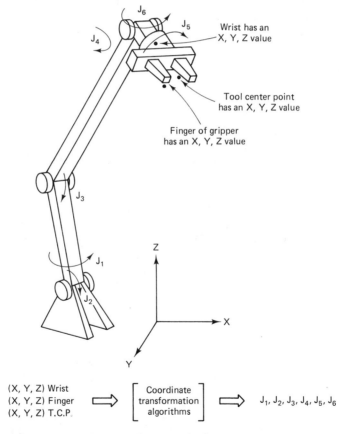

Figure 5-26 The Relationship Between the Cartesian Space Coordinates and the Joint Space Coordinates Is Specified by the Coordinate Transformation Algorithm

The two major differences between this level of control and the joint control level are programming ease and increased control of the tool movement. The program still consists of recorded endpoints for each move, but for the higher level system the trajectory between programmed points can include tool-center-point control. This trajectory control can include straight-line motion, maintenance of tool orientation, tracking of weld seams, and control for grinding. This feature is in contrast to the arc motions that are present in joint control machines. The improved trajectory control is a result of real-time computation of the coordinate transformations. The endpoints are stored in memory as Cartesian, cylindrical, or spherical work space coordinates. The intermediate trajectory points necessary for the desired controlled path are calculated during execution and transformed to the corresponding robot joint angles. Interfacing at this level permits an external computer to deliver program points in the form of X, Y, and Z values. The robot controller converts the X, Y, and Z values into the required joint angles. The external computer does not have to know what type of arm is being driven.

Trajectory Control Level

Interfacing with the controller at this level provides the highest level of trajectory control. The user has the capability of specifying trajectory acceleration and deceleration profiles, trajectory modification based on sensory feedback data, off-line programming which does not require the robot to be moved through the desired path during teaching, and real-time branching to alternate routines. In addition, the interface at this level is arm independent. Arm independence means that the information or data passed to the controller at this interface level need only be related to work cell coordinates and stated in terms of production results desired. The controller has the responsibility to transform the data into the required coordinates and joint angles for the robot arm used in the work cell.

Interfacing remote computers to robot controllers for trajectory control is used in less than one percent of current applications. While it is presently not in demand, the need for a robot control interface will continue to develop as the manufacturing sector moves toward CAM (Computer Aided Manufacturing).

Complex Sensor Interface

The complex sensor interface is used with sensors that require some signal conditioning before the data are transferred to the robot controller. A *complex sensor* is defined as one which requires some type of preprocessor to perform analog-to-digital conversion, scaling, filtering, formatting, analysis, or coordinate transformation on the raw data before they are presented to

Sec. 5-5 Interfaces

the controller. A *complex sensor* communicates with the *complex sensor interface* through either digital or analog signals, but the interface communicates with the robot using only digital signals.

The need for complex sensors and their interfaces results from a blind, dumb, and deaf robot without the sense of touch trying to operate in an unstructured environment. In this type of work cell the robot must have the ability to sense errors in its action and correct trajectories based on sensory measurements. The complex sensor system gathering the data frequently must perform computational operations on the data to

- Convert from analog values to digital values which are compatible with the robot controller
- Detect features or recognize patterns present in the information
- Compare measured data with values previously stored in the robot controller or sensor interface
- Perform coordinate transformations

Data from vision and tactile sensors illustrate the way in which complex computation must be performed on the sensory information before it can be used by the robot controller.

In addition to the computation requirements placed on the complex sensor interface, it must also operate at high speed. The very nature of the job to be accomplished—error correction and trajectory control—requires that the closed-loop response time of the robot and controller be very short. Decisions cannot be made and errors corrected until the data are presented to the robot controller. Delays of more than a few milliseconds do not permit high-performance closed-loop operation.

Bi-directional data communication is another important characteristic of the complex sensor interface. Not only must the interface talk to the controller, but the robot controller must communicate with the interface. In the most complex interfaces the two-way communication includes control signals, serial or parallel data, and addressing information. For example, in vision applications the robot controller signals the vision system when a part needs to be analyzed, and the vision system sends the coordinates and orientation of the part to the controller.

The choice of complex sensor interface is dictated by the type of complex sensory information present. The system may require one sensor per interface if processing speed is important or it may permit one interface to handle many sensors. At present, vision, tactile, and remote positioning tables require the most complex interfaces. Figure 5-27 is a block diagram of a robot controller and complex sensor system. In one of the interface modules, the vision sensor requires a minicomputer with memory for storage of data and vision system programs. In another, a pressure sensor needs an analog-to-digital converter for data transformation. In a third, a digital-tc

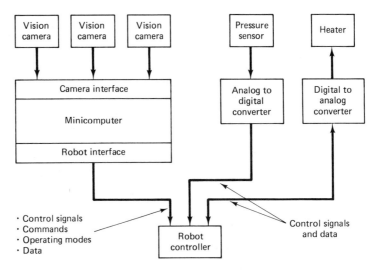

Figure 5-27 Block Diagram of a Complex Sensor Interface

analog converter is used to drive the heating unit. As automated work cells become more complex, the demands on sensor interfacing will increase. Even simple sensors may require complex sensor interfaces to handle the large number of sensors and to determine sensor priority within the system. In some current work cells, for example, over a thousand sensors are required for control of the production line system. With large numbers of sensors, several complex sensor interfaces are required. Figure 5-28 is a block diagram for a robot system with complex and simple sensors interfaced with the robot controller. This type of solution to the system interface problem is called a *star* configuration since all sensors are connected to one central controller. Star configurations are effective for small numbers of sensor connections. The system illustrated in Figure 5-29 is a combination

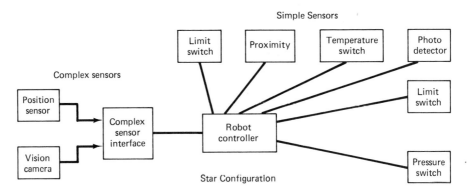

Figure 5-28 Robot Interface in Star Configuration

Sec. 5-5   Interfaces

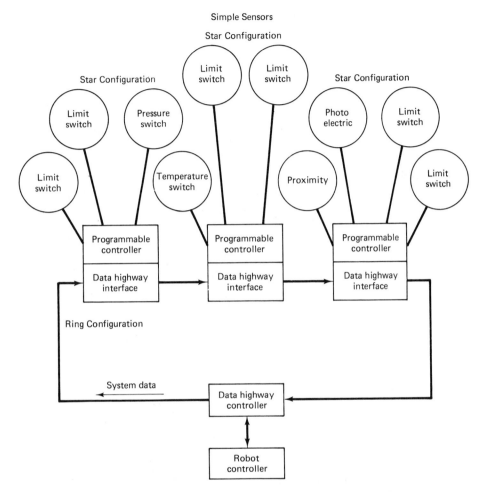

Figure 5-29 Robot Interface in Star and Ring Configuration

of a *star* and a *ring* configuration. The large number of inputs dictated the use of programmable controllers (PCs) as complex sensor interfaces. The sensors connected to the PCs form *star* configurations. More than one PC is used because the total number of inputs exceeds the I/O capacity of a single machine and because distributing the PCs around the work cell reduces the length of runs for sensor wires. The outputs of the PCs are interfaced to the robot controller through a data highway in a *ring* configuration. In this type of configuration the data from every PC pass by the *ring* controller many times a second. The *ring* or data highway controller is programmed to output information to the robot controller in one of three ways: (1) in a fixed sequence, (2) upon request of the robot, or (3) on a priority basis set by the robot system.

5-6 SUMMARY

Sensors make the robot a part of the environment in which it exists. They give the robot information about the work cell which is vital to normal operation. Sensors are used to (1) protect both worker and robot from harm, (2) monitor the production system and work cell operation, (3) analyze product quality, and (4) provide part identification and orientation. All sensors are either contact, noncontact, or process-monitoring devices. Contact sensors include limit switches and artificial skin; noncontact sensors include proximity sensors, photo sensors, and vision systems. Process sensors measure parameters which are part of the production process.

The robot interface permits the controller to communicate with all the systems in the work cell. These systems range from passive switches with no onboard intelligence to complex vision systems with dedicated minicomputers. The controller interface must be compatible with this broad range of signal requirements. All robot interfaces fall into one of the following categories: simple sensor interface, wrist interface, common robot control interface, and complex sensor interface. The simple sensor interface handles discrete signal sensors which do not require signal conditioning. The wrist interface satisfies the need for an interface at the union of the tooling with the tool plate. The common robot control interface permits trajectory control of the robot arm from an external source. Finally, the complex sensor interface provides signal conditioning for complex sensory signals before they are routed to the robot controller.

QUESTIONS

1. What is the basic function of sensors in an automated work cell both with and without a robot present?
2. What are the six basic reasons sensors are used in a work cell?
3. What are the three categories into which all sensors can be grouped?
4. What kinds of sensors are included in the contact sensor group?
5. What is the definition of a limit switch?
6. What are the four basic parts of every limit switch?
7. How does pilot duty differ from electronic duty in the electrical characteristics of limit switches?
8. Using the data sheet for the D4A model general-purpose limit switch, determine the switch part number and lever part number required for rotary operation using fork rollers on the same side, and double-pole–double-throw operation for electronic duty.
9. What is the function of the lamp on a limit switch?
10. Design a trip *dog* to actuate a limit switch in a work cell. The velocity of the trip *dog*

Chap. 5   Questions

is 1.7 feet per second with non-overtravel operation on a low-torque type switch. Draw and dimension the finished design.

11. How does the operation of overtravel *dogs* differ from that of non-overtravel *dogs*?
12. What is the distinction between tactile sensing and simple touch sensing?
13. Describe how tactile sensors operate.
14. What three steps are required for intelligent acquisition of parts?
15. What sensors are included in the noncontact sensor group?
16. Describe the four package configurations now available with proximity sensors.
17. Describe the operation of metallic detection and nonmetallic detection proximity sensors.
18. What three output circuits are available on proximity sensors?
19. Using the data sheets for the TL-X model of proximity sensors, determine the part numbers for a sensor and a sensor controller that satisfy the following conditions: Sensor must be mounted inside a steel block with the sensing surface flush with the block surface. The sensing distance is 4 mm, and the output is normally closed NPN type.
20. List the order of sensitivity for the TL-X model sensor for iron, brass, stainless, and aluminum.
21. What is the sensing distance for a TL-X18M sensor for iron which has a length on one side of 20 mm?
22. Describe the four different types of photoelectric sensors.
23. What is the primary difference between the four types of photoelectric sensors?
24. What type of photoelectric sensor has the greatest range?
25. What type of photoelectric sensor is most accurate when used for range finding?
26. What factors affect photoelectric sensor selection?
27. Select a photoelectric sensor and sensor controller for the following applications. Identify the part numbers of the sensor and controller required.
    (a) A sensor to operate at a distance of 8 cm with high background light. The output should be on when the object is sensed.
    (b) A sensor to operate over a distance of 8 feet to detect when a part breaks the beam. The output should be active when the beam is broken.
    (c) A sensor to count the number of resistors in a parts feeder by using the resistor leads to reflect a beam. The output should be off when the lead is present.
    (d) A sensor to count revolutions of a shaft using marks on a transparent disk mounted to the shaft. The output should be active when the mark is not present.
28. Describe five applications of vision systems currently in use.
29. What are the basic components of a vision system?
30. What two types of cameras are currently used?
31. Describe the operation of a solid-state CCD camera and define the terms *pixel* and *grey scale*.
32. Describe edge detection and clustering.
33. Describe the three commonly used techniques for two-dimensional recognition with vision systems.

34. From the sensors covered in this chapter, list those which would require only a simple sensor interface.
35. What is the definition of an interface?
36. Describe the three interfaces present in the wrist interface.
37. Describe joint space control, coordinate level control, and trajectory level control.
38. List the advantages of standard interfaces for robot controllers.
39. What are the characteristics of sensors using the complex sensor interface?

PROJECTS

1. Make a table which has the six basic reasons for using sensors across the top, and a list of all the sensors used in the work cell in Figure 5-22 down the left side. Indicate the reason for each sensor by placing an X in the appropriate column or columns.
2. Draw a classification tree which includes all the sensors described in this chapter.
3. Design a decision tree to select the best sensor for a given manufacturing problem. Use questions at each branch of the tree that require either a *yes* or *no* response.
4. Write a computer program which will execute the decision tree from project 3.
5. Write a computer program to design limit switch trip *dogs*.

6

ROBOT PROGRAMMING

6-1 INTRODUCTION

Programming languages are the basic communication mechanisms between human beings and the intelligent machines that work for us. Initially, these intelligent machines were the computers themselves, programmed to solve numerous problems in business or scientific areas. More recently, however, computers have been incorporated into other industrial and office machines to increase their efficiency and capability. In spite of this rapid integration of computers into machines of all types, the function of the programming language remains unchanged. Communication continues to be the primary function, but an ever increasing number of sources requires an ever increasing variety of data and information.

The first robots were driven by drum-type sequencers and provided little programming flexibility. Present-day robots are controlled by powerful digital computers, many with multiple processors, which permit a high level of user and machine communications. These computers are responsible for the following system functions:

Manipulation: The control of the motion of all robot joints. This includes position, velocity, and path control of the arm during all programmed motion.

Sensing: The gathering of information from the physical world surrounding the robot. This includes the collection of sensory information and the control of peripheral equipment.

Intelligence: The ability to use information gathered from the work cell to modify system operation or to select various preprogrammed paths.

Data processing: The capability to use data bases and to communicate with other intelligent machines. This includes the capability to keep records, exchange programs, generate reports, and control activity in the work cell.

A manufacturing system with these characteristics would have production flexibility not found in current small-batch systems. If these characteristics are to be realized in small batch-size systems, however, the users must readily be able to specify how a system must operate to produce the product or to perform the task required. Consequently, the need for a suitable programming language which satisfies the system requirements is apparent. This chapter analyzes the current languages used in the robot industry and identifies the kinds of languages which will appear in the future.

6-2 ROBOT LANGUAGE DEVELOPMENT

Traditionally, most robot languages are designed using two techniques. The first approach focuses on developing a language which meets the control needs of the robot arm. When this is satisfied, the language is expanded to include language structures, for example, conditional branching and input/output interfacing. Examples of robot languages developed with this technique are AL, a language developed at Stanford University, and the T3 language developed by Cincinnati Milacron for their industrial robot family. Elegant control of the robot's manipulation and tool path is an advantage of this design. The disadvantage is the absence or inefficient application of the data-processing function.

A second technique for development of a robot language starts with an existing general-purpose computer language such as BASIC or FORTRAN. The existing language is expanded to include the semantics of the robot control system. SAIL, a language developed by the Stanford University Artificial Intelligence Laboratory, fits into this category. A significant advantage of this approach is that the base programming language is well defined, operational, and well documented. The primary disadvantage is that the general-purpose nature of the language forces some design compromises which make the converted version less efficient for use as a robot language.

A third approach includes the design of a new general-purpose language that has the functions necessary to drive the robot arm. This technique permits design trade-offs to be applied in a meaningful way for both the robot application and the users of the language. AML (A Manufacturing Language), the language used by the IBM family of industrial robots, was developed using this technique.

No standards currently exist for robot control languages. There is no interchangeability of programs among manufacturers, and in some cases, only limited interchangeability of programs between models from the same manufacturer. Presently, there are almost as many robot languages as there are robot models. The robot industry has set 1990 as the target date for standardization of the many languages currently in use.

6-3 LANGUAGE CLASSIFICATION

One way to classify the many languages used by robot manufacturers is according to the level at which the user must interact with the system during the programming process. For example, if the user must specify the joint angles for each move, the level of interaction is very low compared to a language which permits the user to specify the motion required in statements such as "Pick up the part." Using this criterion, we can group the current languages into four loosely formulated levels. Of course, overlaps between levels exist and some languages appear to straddle two levels, but the classification process is still valuable. Figure 6-1 shows the four basic levels into which all robot languages are grouped.

Sec. 6-3   Language Classification   169

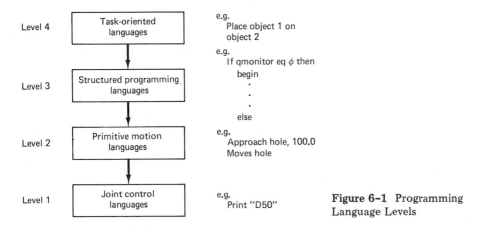

Figure 6-1 Programming Language Levels

Joint Control Languages

Languages at this level concentrate on the physical control of robot motion in terms of joints or axes. The program commands must include the required angular change of rotational joints or the extension lengths of linear actuators. The language usually does not support system or work cell commands, such as INPUT or OUTPUT, which can be incorporated into the programs of higher level robot languages for control of external devices.

This level language requires that the user program in joint space. The term *joint space* means that all the programmed points in the robot's work envelope are expressed as a series of axis positions for all the axes on the arm. Figure 6-2 illustrates a robot arm which has three programmed points. The table included in the figure shows the joint angles which are required

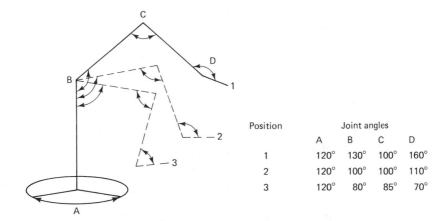

Position	Joint angles			
	A	B	C	D
1	120°	130°	100°	160°
2	120°	100°	100°	110°
3	120°	80°	85°	70°

Figure 6-2 Joint Space Programming

for the robot axes for each programmed point. The advantage of this level of control language is that the software required for control is not difficult or costly to develop. The disadvantages include the lack of integrated work cell commands required for system control, the absence of any Cartesian coordinate values for the programmed points, and severe limitations on path and velocity control.

Examples of this type of language are found most frequently on the many educational robots currently available. The ARMBASIC used on the Minimover 5 robot by Microbot and the RASP language used on the Rhino XR robot are representative of this level. Both robots are jointed-spherical in design, and program commands specify the degree of rotation desired on each axis. In ARMBASIC the move of each axis is expressed by the number of steps which the axis driver, a stepper motor, should rotate. The Rhino uses an optical encoder for feedback from the servomotor on each axis. The Rhino move for each axis is expressed by the number of holes through which the servo and encoder should rotate. Both machines are driven by a microcomputer with an RS-232C interface so the joint space programs for motion are written in any language supported by the microcomputer.

Primitive Motion Languages

Point-to-point primitive motion languages are the most common type used on industrial robots available today. Although the languages included in this group vary widely, they all exhibit the following characteristics:

- A program point is generated by moving the robot to a desired point and depressing a program switch. A sequence of points is saved in this manner, producing a complete program.
- Program editing capability is provided.
- Teaching motion of the robot is controlled by either a teach pendant, terminal, or joy stick.
- The programmed and teaching motion can occur in the Cartesian, cylindrical, or hand coordinate modes.
- Interfacing to work cell equipment is possible. Robot controllers can interact with external signals by using the external signals for control or by signaling external events. Work cell control by the robot system is possible.
- The language permits simple subroutines and branching.

In addition, some of these point-to-point languages permit simple parallel execution using two or more arms in the same work space. The Olivetti language Sigla, for example, permits independent operation of arms with limits and convergence points to prevent collisions. It also permits the execution of several different work files on different arms at the same time with boundaries for each arm set by the anticollision commands in the language.

Some languages in this group have limited coordinate-transformation capabilities. The VAL and RPL languages from Unimation/Westinghouse can define reference frames, invert transformations, and multiply matrices.

The primary advantage of languages at this level is proven performance on the manufacturing floor. The common disadvantage is that the emphasis in programming is still on robot motion rather than on the production problem. In addition, this level does not support the need for off-line programming.

Structured Programming Languages

The structured programming languages offer a major improvement over the primitive motion level. This level contains languages which have the following characteristics:

- A structured control format is present.
- Extensive use of coordinate transformations and reference frames is permitted.
- Complex data structures are supported.
- Improved sensor commands and parallel processing above the previous language level are included.
- State variables are permitted. These are system variables whose value is a function of the state or position of the system. For example, the PAL language uses two state variables named ARM and TOL. The value of ARM changes as the robot arm changes position and orientation. TOL changes as the tooling changes. In the languages Help, MCL, and AL, the user can define the state variables required.
- The format encourages extensive use of branching and subroutines defined by the user.

Not all of the languages listed in this classification exhibit all of the characteristics listed above, but each of them has features which make it appropriate for the structured classification group.

The primary advantage of using a language at this level is the programming advantage gained from the use of transformations. This becomes especially true in complex assembly applications and in support of off-line programming efforts. The major drawback at this level is the increased educational demand placed on the user who must program with transformations in a structured format.

Task-Oriented Languages

The primary function of a task-oriented language is to conceal the use of low-level aids like sensors, branching, and transformations from the user during programming. The user need only be concerned with solving the manufacturing problem. Languages at this level have the following characteristics:

- Programming in natural language is permitted. A natural language command might be "Screw bracket A and bracket B together."
- A plan generation feature allows replanning of robot motion to avoid undesirable situations.
- A world modeling system permits the robot to keep track of objects. This feature provides the systems needed to locate and identify objects, to determine pickup point and orientation of objects, and to move objects relative to one another. In addition, the system can store and use the new relationship between two joined objects.
- The inclusion of collision avoidance permits accident-free motion.

TABLE 6-1 PROGRAMMING LANGUAGES BY LEVEL

Origin	Level 1	Level 2	Level 3	Level 4
Cincinnati Milacron		T3		
IBM		Funky Emily	AML Maple	Autopass
Westinghouse		RPL		
Unimation		VAL		
Sheinman			AL PAL	
Bendix		RCL		
General Electric			Help	
McDonnell Douglas			MCL	
Anorad		Anomatic		
Olivetti		SIGLA		
Automatix		Rail		
Machine Intelligence Corporation		BASIC		
Microbot	Armbasic			
Rhino	RASP			

Sec. 6-5 T3 Language

There are currently no operational languages at this level. Several languages in the developmental stages have some of the necessary characteristics. Included in this group are AUTOPASS (Automatic Programming System for Mechanical Assembly) from IBM, RAPT (Robot Automatically Programmed Tools), and LAMA (Language for Automatic Mechanical Assembly).

A review of industrial applications indicates that robots which execute level two primitive motion languages are the most frequently used in current work cells. Table 6-1 provides a list of languages and the corresponding level of their operation.

6-4 SAMPLE PROGRAMS

An introduction to programming would not be complete without a sample of one current programming technique used by some robots. The objective of this section is to provide an overview of the programming process used on the Cincinnati Milacron T3 robot. In the following paragraphs a sample program is examined in detail with the function of each program line described. The structure and program flow for control of the T3 machine should become apparent. The commands used in this program are briefly defined, but the ones covered by no means exhaust those available to the programmer.

6-5 T3 LANGUAGE

The Cincinnati Milacron T3 language used in this example drives a T3-726 small electric robot through a machine-loading sequence. The robot starts from a HOME position (the starting position for all programs) and moves to a parts feeder to get a part. The gripper acquires a part from the feeder, and the arm moves to a position in front of the machine to be loaded. The robot then loads the part into the machine, retreats from the machine, and starts the production machine cycle. At this point, the arm returns to CYCLE START (a programmed point at the start of the program) and the process is repeated. The program to execute this process is given in Figure 6-3 and includes 36 programmed points.

T3 Procedure

The T3 language is organized around a main program sequence; as many as 255 other sequences can be used as subroutines. The main sequence always starts with a HOME point, includes a CYCLE START point, and ends with a CLOSED PATH command. The HOME point is always the first point in

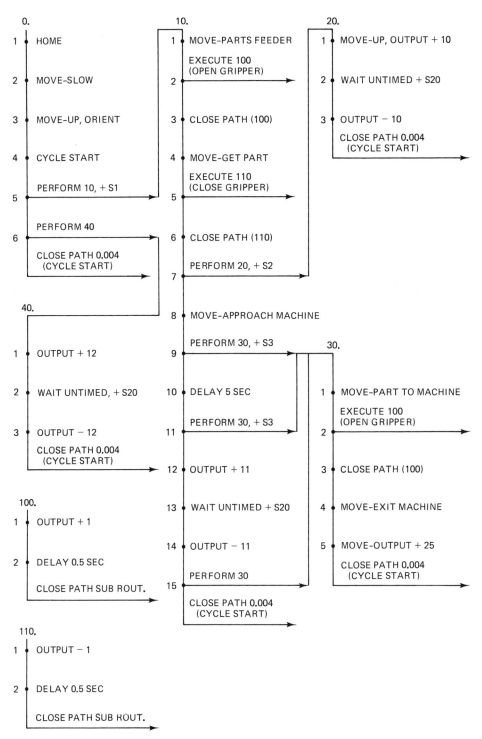

Figure 6-3 Sample Machine Loading Program

Sec. 6-5 T3 Language

the program and is set at start-up to indicate the starting position of the robot arm. Only with the starting position fixed can the robot arm repeat the exact programmed points every time it is operated. Robots are programmed to move through a series of points as they solve a manufacturing problem. The set of programmed points is called a cycle. After completing a cycle of the program, the robot arm will return to the point identified as CYCLE START. CYCLE START is the first programmed point in the robot's solution of a manufacturing problem. The CLOSED PATH command in the main sequence is an unconditional jump back to the CYCLE START point.

Every programmed point must have a *function, velocity, tool center dimension,* and a set of *Cartesian coordinate values* entered during programming. The programmer can either specify the function, velocity, and tool center dimension at each point before pressing the program button or simply use the current values from the last programmed point. The coordinate values are automatically recorded from the position of the arm when the point is programmed. Also, every programmed point is identified by a sequence number and a programmed point number. Figure 6-4 illustrates the program point numbering system used in the T3 controllers. In the sample program, Figure 6-3, the CYCLE START point is 0.004, which means it is point 4 in the 0 sequence. The sequence number and programmed point number are separated by a period. The HOME programmed point is 0.001 in every program. The main sequence is always designated as the 0 sequence. With 256 possible sequences and 999 possible programmed points in each sequence, the total number of programmed points possible with the controller is 255,744.

Sequences 1 through 255 can be called from the main sequence in the same manner that subroutines are called in a high-level language like BASIC. In addition, a sequence can be called from within another sequence. The T3

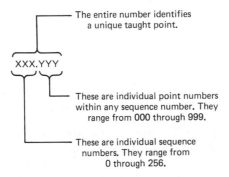

0.329 = Point 329 in sequence 0 (Mainline)
26.001 = Point 1 in sequence 26
179.022 = Point 22 in sequence 179

Figure 6-4 Program Numbering System (*Courtesy of Cincinnati Milacron*)

language supports either conditional or unconditional branches to any sequence. The conditional branch uses the state of an external signal, an external interrupt, or the value of a variable to determine if the condition for branching has been satisfied. The unconditional branch is performed whenever the command is encountered and is similar to a GOTO command in BASIC.

All sequences must be terminated in a CLOSED PATH command. There are four different CLOSED PATH commands that can be used with either a conditional or an unconditional call. Two of the commands are:

>CLOSED PATH, ABSOLUTE, FIXED POINT

>CLOSED PATH, ABSOLUTE, SUBROUTINE

The ABSOLUTE, FIXED POINT is similar to the GOTO command in BASIC since it requires that the branch point be a specific programmed point. For example, the CLOSED PATH command at the end of sequence 20 in the program in Figure 6-3 is an ABSOLUTE, FIXED POINT type of CLOSED PATH. The ABSOLUTE, SUBROUTINE command is similar to the RETURN command which is used at the end of subroutines in BASIC. This sequence terminator will return program control to the programmed point after the calling point when the called sequence is completed. In the sample program, sequences 10 and 100 have this type of subroutine structure. Program point 10.002 calls sequence 100. When sequence 100 is complete, the CLOSED PATH, SUBROUTINE command at the end returns control to sequence 10. The next program point to be executed is 10.003.

T3 Commands

In addition to the commands described previously, the following commands are used in the sample program in Figure 6-3:

>Conditional branch: PERFORM, (seq. no.), (input)

The conditional branch command, PERFORM, jumps to the first programmed point in the sequence specified (seq. no.) if the input condition is satisfied. The input is identified by a positive or a negative sign, the letter *s*, and the input channel number. A positive sign requires that the input channel be active or *on* for the condition to be satisfied.

>Unconditional branch: PERFORM, (seq. no.)

The unconditional branch command, PERFORM, jumps to the first programmed point in the sequence specified (seq. no.) and executes the programmed points in the new sequence starting with the first.

>Unconditional branch: EXECUTE, (seq. no.)

Sec. 6-5   T3 Language

The unconditional branch command, EXECUTE, jumps to the first programmed point in the sequence specified (seq. no.) and executes the functions programmed at each programmed point in the called sequence but not the moves.

Wait: WAIT UNTIMED, (input)

The wait command, WAIT UNTIMED, causes the robot to halt at this programmed position and wait for an input condition which will trigger a restart of the programmed routine. The input will follow the same syntax used on the conditional jumps.

Wait: DELAY, (number of seconds)

The wait command, DELAY, causes the robot to remain motionless at this point for the amount of time specified.

Output: OUTPUT, (sign and channel number)

The I/O command, OUTPUT, permits robot control of equipment external to the controller. When executed, this programmed point turns *on* the specified channel if the sign is positive. A negative sign turns the channel *off*.

Input and Output Signals

Input signals from the work cell are used to control the operation of the robot by providing the information needed to make program decisions. In the T3 language inputs are designated by the letter *s* followed by the input channel of the signal. The sign preceding the letter *s* indicates the anticipated condition on the channel. A positive sign requires an active or *on* input condition, and a negative sign requires an *off* condition. The inputs for the sample program are defined as follows:

+ s 1   Channel 1 is a signal from the parts feeder. An *on* condition indicates that parts are ready for pickup by the robot. Channel 1 will be *off* when the feeder is empty.

+ s 2   Channel 2 is a signal from the robot gripper. An *on* condition indicates that the part is not oriented properly in the gripper. If the part is gripped properly channel 2 will be *off*.

+ s 3   Channel 3 is a signal indicating the machine is ready for the robot to insert a new part. An *on* condition indicates that the machine cycle is complete, the previous part has been ejected, and a new part is needed. If the machine cycle is not completed or if the part did not eject, then the signal on channel 3 will be *off*.

+ s 20 Channel 20 is a signal indicating the operator has corrected the problem which caused the work cell to halt. Pushing the restart button forces channel 20 to an *on* condition.

Output signals from the controller operate external equipment under the program control of the robot. The following outputs are interfaced to the indicated equipment in the work cell:

Output 1 The gripper solenoid is connected to this output. An active output or *on* condition causes the gripper to open. If a negative output is indicated, the *off* condition causes the gripper to close.

Output 10 Output channel 10 is connected to an indicator lamp labeled **PART NOT ORIENTED**. If this output is active the lamp is *on*. The lamp is turned *off* when the channel is turned *off* with a negative output.

Output 11 Output channel 11 is connected to an indicator lamp labeled **MACHINE CYCLE INCOMPLETE**. Again, a positive output turns it *on* and a negative output turns it *off*.

Output 12 Output channel 12 is connected to an indicator lamp labeled **PARTS FEEDER EMPTY**. The operation is identical to 10 and 11.

Output 25: Output channel 25 starts the production machine cycle. An active output turns the machine *on*.

Sequence Functions

Each sequence illustrated in the sample program in Figure 6-3 has a different function in the manufacturing process performed by the robot. The function of each sequence is described as follows:

Sequence

0 **Main sequence:** Moves the robot to cycle start and checks the status of the parts feeder.

10 **Get Part sequence:** Moves the robot to the parts feeder, calls the open gripper sequence, gets the part, calls the close gripper sequence, checks the part orientation, moves to the production machine, and tests the readiness of the production machine for a new part. Sounds alarm if production machine is not ready in five seconds.

20 **Alarm sequence:** Moves the robot to a safe position, turns on the **PART NOT ORIENTED** indicator lamp, and waits for an

	operator to correct the problem. When corrected the sequence turns *off* the alarm.
30	**Insert Part sequence:** Moves the robot to insert the part in the production machine and initiates the cycle of the production machine.
40	**Alarm sequence:** Turns on the **PARTS FEEDER EMPTY** indicator lamp and waits for operator intervention. When corrected the lamp is turned *off*.
100	**Gripper Open sequence:** Opens the gripper.
110	**Gripper Close sequence:** Closes the gripper.

6-6 PROGRAM ANALYSIS

To provide an overview of robot programming via the T3 language, we have taken the sample program in Figure 6-3 and analyzed it in detail. The function assigned to each programmed point is identified. In many cases the programmed point requires only that the arm make a move to a new position, and in these cases the function will be a NOP or No OPeration. The description of each program line follows:

0.001	HOME position
0.002	NOP: Arm moves out from HOME at a slow velocity.
0.003	NOP: Arm moves up and puts the tooling in the correct orientation.
0.004	CYCLE START

WORK CELL CONDITION — PARTS FEEDER FULL

0.005	PERFORM 10, +s1: The program branches to sequence 10 if the signal on input channel 1 is *on*, which means that there are parts in the parts feeder. If the signal on channel 1 is *off*, the program executes the next line in sequence 0 (0.006) since the parts feeder is empty.
10.001	NOP: Since the branch has been executed, the arm moves to the parts feeder.
10.002	EXECUTE 100: The program branches unconditionally to sequence 100. Since it is an EXECUTE command, no arm moves occur.
100.001	OUTPUT +1: The gripper solenoid on channel 1 output is turned *on* and the gripper opens.

100.002	DELAY .5 sec: The arm delays .5 seconds for the gripper to open. No arm motion occurs.
—	CLOSED PATH SUBROUTINE: Program control is returned to sequence 10 at the programmed point (10.003) following the branch point (10.002).
10.003	NOP
10.004	NOP: Arm moves to grasp part in feeder.
10.005	EXECUTE 110: The program branches unconditionally to sequence 110. No arm move occurs since it is an EXECUTE command.
110.001	OUTPUT −1: The gripper solenoid on channel 1 output is turned *off* and the gripper closes.
110.002	DELAY .5 sec: The arm delays .5 seconds for the gripper to close. No arm move occurs.
—	CLOSED PATH SUBROUTINE: Program control is returned to sequence 10 at the programmed point (10.006) following the branch point (10.005).
10.006	NOP

WORK CELL CONDITION − PART ORIENTED IN GRIPPER

10.007   PERFORM 20, +s2: The input to channel 2 is *off* since the part is oriented in the gripper. Because the condition for the branch is not satisfied, it will not occur. Branch will only occur if the part is not oriented.

10.008   NOP: Arm moves to production machine.

WORK CELL CONDITION − PRODUCTION MACHINE BUSY

10.009   PERFORM 30, +s3: The input on channel 3 is *off* since the production machine is still busy with the last part. The condition for the branch is not satisfied so the branch will not occur.

10.010   DELAY 5 sec: The arm delays 5 seconds for the production machine to finish the cycle. No arm movement occurs.

WORK CELL CONDITION − PRODUCTION MACHINE READY

10.011   PERFORM 30, +s3: The input on channel 3 is *on*, since the production machine is finished with the previous part. The condition for branching is satisfied so a branch to sequence 30 will occur. If the production machine was still not fin-

Sec. 6-6 Program Analysis

ished, then the next step in sequence 10 would have been executed.

30.001   NOP: Arm moves the part to the machine.

30.002   EXECUTE 100: The subroutine at sequence 100.001 is called. Operation is the same as at 10.002 except the return point is step 30.003. When the gripper is opened, the part is placed in the machine.

100.001   OUTPUT +1: The gripper solenoid on channel 1 output is turned *on* and the gripper opens.

100.002   DELAY .5 sec: The arm delays .5 seconds for the gripper to open. No arm motion occurs.

—   CLOSED PATH SUBROUTINE: Program control is returned to sequence 30 at the programmed point (30.003) following the branch point (30.002).

30.003   NOP

30.004   NOP: Gripper exits the machine.

30.005   OUTPUT +25: The machine cycle start control is interfaced to robot output 25. The machine is started when the output is activated or turned *on* with this program step. In addition, the arm starts the move back to the CYCLE START position.

—   CLOSED PATH 0.004: Program control is returned to the main sequence at step 0.004 (CYCLE START), and the robot cycle repeats itself.

The sequence steps just described in this program represent the ideal robot motion and work cell operation, that is, without any malfunctions. However, sensors connected to robot I/O channels one, two, and three do check three malfunctions in the work cell. The operation of these malfunction sequences are described as follows:

WORK CELL CONDITION — PARTS FEEDER EMPTY

10.005   PERFORM 10, +s1: The program does not branch to sequence 10 because the input on channel 1 is *off* as a result of no parts in the feeder. If no branch occurs then the next step in the sequence is executed.

10.006   PERFORM 40: This is an unconditional branch to sequence 40.

40.001   OUTPUT +12: The warning light labeled **PARTS FEEDER EMPTY** on output 12 is turned *on* by this program step.

40.002   WAIT UNTIMED +s20: The robot arm enters a wait state which will last until the input on channel 20 is activated or turned *on*. The restart button connected to channel 20 is pressed by an operator after the parts feeder is refilled.

40.003   OUTPUT −12: The warning light labeled **PARTS FEEDER EMPTY** on output 12 is turned off by this program step.

—   CLOSED PATH 0.004: Program control is returned to the main sequence at step 0.004 (CYCLE START) and the robot cycle repeats itself.

Two alternate approaches can be used for the alarm sequence just described. In the first, the branch instruction, PERFORM 40, in step 10.006 could be changed to an EXECUTE 40, since there are no moves programmed in the alarm sequence. In the second, the entire alarm subroutine, sequence 40, could be included in the main sequence as steps starting at 0.006. Each approach has advantages and disadvantages which must be considered before selecting the specific programming sequence.

WORK CELL CONDITION — PART NOT ORIENTED

10.007   PERFORM 20, +s2: When the part picked up by the robot is not correctly oriented for loading into the production machine, the program branches to sequence 20. This condition is detected by a sensor and turns channel 2 *on* when the part is not oriented.

20.001   OUTPUT +10: This step turns *on* the warning light labeled **PART NOT ORIENTED** on channel 10. In addition, the arm is moved out to a safe position.

20.002   WAIT UNTIMED +s20: The robot waits for the operator to clear up the orientation problem. The robot cycle will resume only when the operator presses the restart button connected to input 20. An *on* condition at input 20 will restart the robot.

20.003   OUTPUT −10: The warning light turned *on* in step 20.001 with a positive output is now turned *off* with a negative output.

—   CLOSED PATH 0.004: Program control is returned to the main program at step 0.004 (CYCLE START), and the robot cycle starts over.

WORK CELL CONDITION — PRODUCTION MACHINE BUSY

10.011   PERFORM 30, +s3: Sequence 30, loading the production

QUESTIONS

1. Describe the four system functions performed by the computer in high-technology robots.
2. Describe the three techniques used to develop a robot language, and list two languages which were developed by each method.
3. Describe the difference between joint space programmed points and Cartesian space programmed points.
4. What are the four levels of robot languages?
5. What are the advantages and disadvantages of languages at each level?
6. What are the common characteristics of languages at each level?
7. What four program parameters must be entered at every programmed point in the Cincinnati Milacron T3 language?
8. What is the difference between sequence 0 and sequences 1 through 255 in the T3 language?
9. What is the difference between the fixed point and the subroutine types of closed-path commands?
10. What branching options are available in the T3 language?
11. What do the positive and negative signs on the input and output signals indicate?

PROJECTS

1. Modify the program in Figure 6-3 to load two production machines instead of one. Use additional inputs and outputs as required.
2. Write a program for the forging operation illustrated in Figure 5-22.

machine, is not performed unless the previous part has been ejected. With input 3 *off*, the branch is not performed, and a warning program routine is executed.

10.012   OUTPUT +11: This step turns *on* the warning light labeled **MACHINE CYCLE INCOMPLETE** on channel 11.

10.013   WAIT UNTIMED +s20: The robot enters a waiting period until the operator clears up the problem in the production cycle. The cycle is restarted with the restart button which is connected to input channel 20.

10.014   OUTPUT −11: The warning light labeled **MACHINE CYCLE INCOMPLETE** on channel 11 is turned *off* by this step.

10.015   PERFORM 30: The unconditional branch to sequence 30 is executed in this step. With the machine cleared, the robot can load the next part into the machine. The operation of sequence 30 is described earlier.

The CLOSED PATH 0.004 command which appears at the end of the main sequence and at the end of sequence 10 will never be executed. The unconditional branch which precedes these commands will never permit program flow to reach this last statement. The commands must be present, however, since every sequence must be terminated with a CLOSED PATH statement.

6-7 SUMMARY

The computers used in robot systems are responsible for four functions: (1) machine manipulation, (2) sensing, (3) logical decision making, and (4) data processing. The programming languages currently used to execute these four functions have no common format or standard form. There are as many languages as there are robot manufacturers. However, the many languages were developed using three techniques. The first method starts with the design of a pure manipulator control language and adds the other high-level language features as demand requires. The second adopts an existing high-level language and augments it with the manipulator commands required for robot operation. The last technique includes the design of a high-level language for robot control and support of other external data processing functions.

The many current languages are classified into four groups based on the level at which the user must interact with the system during the programming process. The four groups are joint control languages, primitive motion languages, structured programming languages, and task-oriented languages. At present, the most frequently used languages are in the primitive motion category. Table 6-1 provides a list of languages and their level of operation.

7

SAFETY

7-1 INTRODUCTION

In the past, we have made the acquaintance of robots through science fiction writers. The robots of Isaac Asimov, in his writings of the 1940s and 1950s, come especially to mind. Yet even these fictional machines were subject to Asimov's Three Laws of Robotics:

1. A robot must not harm a human being, nor through inaction allow one to come to harm.
2. A robot must always obey human beings, unless that is in conflict with the first law.
3. A robot must protect itself from harm, unless that is in conflict with the first and second laws.

These laws are still valid today, and no investigation of robotics would be complete without considering safety in robot usage and applications.

The necessity of safety features in the robot work cell can be illustrated by the following analogy: Today's robot can be compared to a man with one arm tied behind his back, both feet nailed to the floor, blindfolded, wearing ear muffs, and gagged. Added to this is the robot's inability to make independent decisions. As a result, the robot may obey incorrect instructions—or fail—because of a malfunction, and may thereby cause damage to itself or nearby equipment, or injury to nearby persons.

Worker safety is the most important concern; of secondary importance, however, is the fact that damaged equipment causes downtime, which decreases production. It is, therefore, the responsibility of work cell designers to consider measures to diminish the possibility of accidents to both hardware and humans. In addition, it is the responsibility of the plant safety director to orient workers to the potential hazards of robotics installations.

7-2 OSHA

The Occupational Safety and Health Act of 1970 represents the foundation for industrial safety standards throughout the nation. Part 1910 of OSHA is particularly helpful as a guideline for safety standards during the design and implementation stages of robot applications. OSHA standards may be viewed from the following two perspectives: (1) safety in the work cell and (2) safety in the robotic application. Presently there are no written regulations that detail the safety requirements for robots; however, Subpart O, Section 1910.212 details general requirements for safety precautions around machinery. The major emphasis of this section is on machine guarding, which protects the operator and others from danger at the point of operation (the area where the work is actually taking place).

For nonrobotic fixed machinery, OSHA requires a fixture that prevents operators from exposing any part of their body to the dangers of the operation. However, the addition of machine guards for human safety within the robot cell may impose severe limitations on the robot's mobility and its access to all parts of the work envelope. One way to view machine guarding is to consider the entire work cell as the point of operation; then the designer can plan methods to protect the workers from such hazards as pinch points, shear points, and collisions within the entire robot work envelope.

A second way to view safety within the work cell is from the perspective of the actual robot production application (welding, material handling, grinding, and so on). Since most current production applications are covered in detail by OSHA regulations, the rules governing a particular application should be reviewed prior to the design of the robot work cell. Following is a list of pertinent sections:

Subpart N	Section 1910.176	Handling Materials—General
Subpart O	Section 1910.215	Abrasive Wheel Machinery
Subpart O	Section 1910.216	Mills and Calenders
Subpart O	Section 1910.217	Mechanical Power Presses
Subpart O	Section 1910.218	Forging
Subpart Q	Section 1910.252	Welding, Cutting, and Brazing

In addition, special OSHA requirements may be requested at the time of inspection. For example, a recently installed robot application in the United States was approved by OSHA only when the following criteria were met: erection of a two-rail pipe fence around the robot work area, with the top rail 54 inches from the floor and the lower rail midway between the top rail and the floor.

7-3 GENERAL PERSONNEL SAFETY

A robotics installation is still a new occurrence in most factories, and the robot itself can be a congregating point for general plant personnel. It is important, therefore, that some type of protective barrier be placed around the work envelope to prevent unauthorized workers from getting too close to the robot work cell and to restrict traffic flow in the area.

Both physical barriers and electronic detectors can be employed. Some examples of physical barriers are:

- Painted lines on the floor to mark the limits of the work envelope
- Chains and guard posts
- Safety rails

Sec. 7-4   Operator and Maintenance Personnel Safety   189

- Wire mesh fencing
- Equipment within the work cell

Painted or taped lines should always be the first step, but never the last, in defining the work area, since these are really a psychological, not a physical barrier. The fourth choice, the wire mesh fence, offers the most complete protection, including a shield against parts which may be accidentally thrown from the gripper during material handling operations. The last type of barrier, the cell equipment itself, when combined with guard rails or fences, provides safety protection at reduced cost. An example of this technique is illustrated in Figure 5-22.

A gate of some type in any physical barrier is required to admit authorized operators and maintenance workers to the robot work area. An electrical interlock between the gate mechanism and the robot control should be installed, so that when the gate is opened the robot ceases motion. However, this cessation should not be confused with the initiation of an emergency stop. Rather, the emergency signal should be avoided whenever possible, because such shutdowns can cause damage to the end effector of a hydraulic robot, for example, when the arm drops after hydraulic power is interrupted. Motion should be halted safely through the desired interrupt process, then the system can be restarted from the controller after the work cell is clear and the gate is closed.

In addition to physical barriers, protection can also be provided by detectors such as electronic curtains, motion detectors, and pressure-sensitive floor pads. The electronic curtains can use visible light, infrared light, or lasers with two or more beams. Laser light can be reflected by mirrors to form a multi-beam light curtain. The second type, motion detectors, sense the presence of a person in the work area using either microwave or ultrasonic signals reflected by the intruder back to the sensor's receiver. The final type of detector uses floor pads with imbedded switches which are activated by the weight of the intruder. A typical robotic installation might include a combination of barriers and protection devices.

7-4 OPERATOR AND MAINTENANCE PERSONNEL SAFETY

Robot application designers need to give even more careful thought to persons who work within the confines of the robot work space. One of the primary considerations in planning for the safety of operator and maintenance personnel is choosing the best method to buffer human beings from the robot. The cell operator should of course be protected by the cell design; however, this protection exists only during normal work cycles. In actual fact, programmers and maintenance workers must sometimes work within the cell during abnormal conditions while the robot is operating. Protecting personnel under these conditions creates the most difficult accident-control problem. Ninety percent of all robot-related accidents occur during maintenance and programming of the robot.

Since safety consciousness begins with the individual worker, the following guidelines should be emphasized to each worker:

Rule 1: Respect the robot. Don't take the robot for granted or make an assumption about the next movement the arm will take.

Rule 2: Know where the closest emergency stop button is at all times. Remember, hydraulic robots which have been stopped under emergency conditions will droop and could come to rest on the floor and damage the gripper.

Rule 3: Avoid pinch points when programming or working in the work cell.

Rule 4: Know the robot. Pay attention to unusual noises and vibrations from the machinery.

Rule 5: Never permit untrained personnel to operate the robot system.

Rule 6: Don't adjust the robot control mechanism without the proper training and supervision.

7-5 SUMMARY

Safety is an important factor in robotic cell design, from the beginning of the design to the implementation of the process. Specific areas of safety awareness in a robotic installation include OSHA requirements, types of barriers for worker protection, general personnel safety, and operator and maintenance personnel safety.

QUESTIONS

1. What are Asimov's three laws of robotics?
2. What is the greatest concern in the work cell from a safety standpoint?
3. What are current OSHA standards for work cells and applications?
4. What is the primary method used to make work cells safe for general personnel?
5. When do most robot-related accidents occur?
6. What rules should be followed to protect the operator and maintenance personnel?

8

HUMAN INTERFACE: OPERATOR TRAINING, ACCEPTANCE, PROBLEMS

8-1 INTRODUCTION

Automated processes are most frequently implemented because of an anticipated increase in productivity or product quality. The impact of automation is felt by almost every segment of society, from the suppliers of automated hardware to the workers who will be operating the system, to the employees displaced by automation, to the consumers of the products of automation. The planning process for implementing an automated system may schedule sufficient time and resources to solve the engineering problems involved but often overlooks the need for an equal emphasis on the human interface which is present in all such systems.

Three distinct training activities should be implemented by every industrial plant that considers introducing robots to the production process:

1. General employee awareness training with an overview of the new types of automation, especially robots
2. Operation and programming training programs for all personnel involved in the production process
3. Maintenance training programs for all personnel who will be responsible for maintaining the robot system

The time frame for the three programs is also critical. Plans should be made to have the appropriate phases of the training occur before, during, and after the implementation process.

8-2 GENERAL TRAINING

"Robots are just another form of automation" is a statement that has often been made by those responsible for developing automated manufacturing systems. The statement can be supported from the hardware standpoint without difficulty, but because of the way robots are perceived by both management and labor, training techniques for robot projects must be especially well thought out. Although the difference between robotic automation and automation in general is more psychological than material, it is a potent force in the minds of both employees and management. The difference stems from the media attention which robots have drawn and from the association of industrial robots with the walking, talking machines of past and present science-fiction films.

For workers, robots create a higher level of anxiety over possible job loss than do other automated processes, and they also encounter greater resistance at the management level. Management resistance starts at the first-line supervisor level and extends up through varying levels of middle management. Therefore, a well-structured general training program is necessary to deal with personnel's concerns about robot automation.

General Training Program

The general training program to prepare all levels of labor and management for the implementation of robots should include the following procedures:

- An overview of current automation practices used by industry in general and competitive industry in particular
- A thorough explanation of the need for robot automation based on production and marketing data
- A complete description of robot automation including what robots are and what they can and cannot do
- A comprehensive statement of company policy regarding jobs eliminated or changed as a result of automation
- A detailed plan describing how all employees affected by the automation will be retrained
- A clear statement of support from the highest level of management for including automation in any productivity improvement process

The audience receiving this general training is extremely varied, both in the depth of information needed and the emphasis on information delivered. The management level, for example, would profit from a complete overview of current and future robot automation for the industry and from training in the specific skills required to implement robots. The production worker, on the other hand, would be most interested in the company's plans for workers displaced as a result of automation.

The timing of the training is also important. Starting well in advance of the first robot implementation is important, but even those companies that currently use robots in production would profit from a general training program before the next installation. Continued information exchange during and after installation is important to keep everyone involved in the current project or future projects up to date.

Training techniques for general audiences are as varied as the groups of listeners they address. General information on robot automation may be a topic for an employee newsletter or company paper. Bulletin boards, quality circle groups, shop and department meetings, and seminars can also be effective training media.

The rationale for general training programs of this nature aimed at increasing understanding about robot automation among all employees is simply that successful implementation requires team effort, and informed workers are better team players.

8-3 OPERATOR TRAINING

The term *robot operator* remains undefined. In one case the human operator may simply activate the work cell at the start of the shift, but in other situations the person works side by side with the robot. The list of responsibilities for a robot operator may include programming, program loading, work cell support, and routine system checkout. Currently, support engineering groups do most of the programming and system checkout, and operators perform that portion of their former job which the robot automated work cell cannot do.

Operator training related to production in the work cell generally takes place in the plant facilities. Frequently, on initial robot applications the operator accompanies engineering and production representatives to robot training classes held at the robot vendor's training facilities.

Most robot vendors recommend that training at their facilities begin two to four weeks before the machine arrives.

8-4 MAINTENANCE TRAINING

In order to understand the maintenance requirements for robots it is necessary to review the type of hardware which will be present in the work cell. Robots currently available are powered by hydraulics, pneumatics, or electric drives. The electric drives are primarily the dc servo type, but ac servomotors are now being introduced. The mechanical mechanism itself has the standard gear, pulley, belt and cam linkages, and drives found on automated machines used in manufacturing. Every robot system has a controller, which is either a special-purpose computer or a programmable controller. The special-purpose computers use either microprocessor technology or minicomputer technology as the central processing unit. The robot work cell includes all the manufacturing hardware required for production, one or more robots, sensors, and material handling equipment.

Maintenance training on the robot, manufacturing hardware, and other components of the work cell has generally become a routine procedure. Robot vendors, like most manufacturers of production hardware, have one- and two-week training programs designed to train maintenance personnel. The training includes preventive maintenance and troubleshooting of hardware failures. Using the built-in diagnostics in the system, an operator can isolate most of the problems to a printed circuit board so that troubleshooting at the board level is all that is required. If a failure cannot be located, the vendors have field service personnel who will come to a facility and do in-plant repair.

Maintenance training at the system level rests squarely on the shoulders of the end user. The work cell typically has equipment from five or more vendors, and each piece of hardware may contain an electronic control system or computer. The interfacing of the hardware to achieve coordinated operation is usually the responsibility of corporate engineering or a systems engineering company. Because of the unique nature of each package, diagnostic tests are usually not available for use by maintenance personnel, and system failures frequently require troubleshooting to the component level. Development of an effective maintenance training program for the system level in a manufacturing work cell presents the stiffest challenge in the training area.

8-5 RESISTANCE

Neal Clapp, an industrial consultant who worked with General Electric Company on reducing resistance to robots, offers the following three guidelines to help overcome management and worker resistance:

1. Organizations may not install robots to the economic, social, and physical detriment of workers or management.
2. Organizations may not install robots through devious or closed strategies which reflect distrust or disregard for the work force. For surely they will fulfill their own prophecy.
3. Organizations may only install robots on those tasks which, while currently performed by man, are tasks where the man is like a robot, not the robot like a man.

Many manufacturers may find it difficult to adhere to the ethics expressed in these laws, but maintaining some degree of openness, as in number two, will certainly aid in the process of robot implementation.

Resistance to robot automation at the management level stems from the following concerns:

- Concern by middle managers that an unsuccessful robot implementation will endanger their future promotions
- Fear by first-line supervisors that robot automation will cause additional employee problems and that they will not have management's support in solving them
- Concern by first-line supervisors that production output will drop as robot automation is brought up to speed, and they will be held accountable
- Conflicts over responsibility for the project, along with jealousy between manufacturing units and disagreements over distribution of resources

Resistance to robot automation at the labor and management levels arises basically from a fear of the unknown. This could be fear of losing a job, fear of being transferred, or fear of the demands of retraining. Much of the resistance resulting from these fears and those of managers can, however, be neutralized by an effective training program at the three levels defined earlier.

8-6 ORGANIZED LABOR

Since organized labor believes that new union members don't want to do many of the tasks that are dirty, dangerous, or dull, it will officially welcome robots in many areas. In addition, labor realizes that increased productivity is necessary. However, organized labor stands firm on two demands: (1) displacement of workers must be gradual and (2) labor should get a percentage of the benefits which derive from robot automation.

Labor acknowledges robot technology, but is reluctant to give management a free hand in its implementation. Unions still fear large-scale displacement of blue-collar workers as a result of automation. Unions are based on the principle of solidarity among workers, and as the ranks decline so does the union's operating revenue. Job security and training will be the major issues in most contract negotiations in the 1980s and beyond. In the coming decades union membership drives will concentrate on the growing number of white-collar workers, especially those involved with automated equipment.

8-7 SUMMARY

Resistance to automation can come from many directions. The degree to which this resistance is neutralized depends on the time and resources which are budgeted for development of the human interface in all automation systems.

QUESTIONS

1. What are the three training activities required in every industrial plant planning to use robots?
2. What elements should be present in a general training program?
3. What is the primary focus of a general training program for each level within an industrial organization?
4. What are the major elements of an operator training program?
5. What complicates maintenance training procedures on automated systems?

6. What is the difference between maintenance at the system level and maintenance at the machine level?
7. What are Neal Clapp's three laws for reducing resistance to automation?
8. What causes resistance to robot automation at the management level?
9. What is organized labor's view of automation and robotics?

APPENDIX A

OMRON GENERAL-PURPOSE LIMIT SWITCH

Cat. No. C03-E3-2   Model **D4A**

Heavy Duty Plug-in Limit Switch — Watertight and Oiltight

■ FEATURES

- Convenient front-mount plug-in construction.
- High-reliability design.
- Wide variety of operating heads.
- Meets NEMA types 1, 2, 3, 3R, 4, 4X, 5, 6, 12 and 13.
- Certainty of dual Viton® sealing.
- Easy installation and maintenance.
- Reduces inventory.
- Low temperature type (−40 to +158° F) is also available.

■ AVAILABLE TYPES

- **SWITCH UNITS** (D4A-□□□□: Add the appropriate code when placing your order, e.g., D4A-1101.)

Operating head	Circuit	SPDT DOUBLE BREAK				DPDT DOUBLE BREAK	
	Indicator lamp	Without lamp		With lamp		Without lamp	
	Contact rating	Pilot duty	Electronic duty	Pilot duty	Electronic duty	Pilot duty	Electronic duty
Side rotary type*	Standard	-1101	-1201	-1301	-1401	-2501	-2601
	High precision	-1102	-1202	-1302	-1402	-2502	-2602
	Low torque	-1103	-1203	-1303	-1403	-2503	-2603
	High precision/Low torque	-1104	-1204	-1304	-1404	-2504	-2604
	Maintained	-1105	-1205	-1305	-1405	-2505	-2605
	Sequential operating	—	—	—	—	-2717	-2817
	Center neutral operating	—	—	—	—	-2918	-2018
Side plunger type	Plain	-1106	-1206	-1306	-1406	-2506	-2606
	Vertical roller	-1107-V	-1207-V	-1307-V	-1407-V	-2507-V	-2607-V
	Horizontal roller	-1107-H	-1207-H	-1307-H	-1407-H	-2507-H	-2607-H
	Adjustable	-1108	-1208	-1308	-1408	-2508	-2608
Top plunger type	Plain	-1109	-1209	-1309	-1409	-2509	-2609
	Roller	-1110	-1210	-1310	-1410	-2510	-2610
	Adjustable	-1111	-1211	-1311	-1411	-2511	-2611
Wobble lever type	Spring wire	-1112	-1212	-1312	-1412	-2512	-2612
	Plastic rod	-1114	-1214	-1314	-1414	-2514	-2614
	Cat whisker	-1115	-1215	-1315	-1415	-2515	-2615
	Coil spring	-1116	-1216	-1316	-1416	-2516	-2616

- **LEVERS FOR SIDE ROTARY HEAD**

	.75 dia./.31 width	.88 dia./.59 width	
Standard-roller front mounted	●		D4A-A00
		●	D4A-B06
Standard-roller back mounted		●	D4A-A10
Offset-roller front mounted		●	D4A-A20
Offset-roller back mounted		●	D4A-A30
Adjustable-roller			D4A-C00
Rod-adjustable			D4A-D00
Fork-rollers on opposite side	●		D4A-E00
Fork-rollers on opposite side		●	D4A-E10
Fork-rollers on same side			D4A-E20

NOTES: 1. * When placing your order for a side rotary type switch, order the lever with the switch.
2. With the standard types, only the switch body and shaft parts are Viton® seated and all other parts are sealed with nitrile-butadiene rubber (NBR). If the Viton® sealed type (with all parts sealed with Viton® rubber) is required, add suffix code "F" to the desired part number shown on the left (e.g., D4A-1101-F).
3. If the low temperature type is required, add suffix code "T" to the part number shown on the left (e.g., D4A-2501-T).
4. To order switch bodies, receptacles and operating heads separately, refer to "REPLACEMENT PARTS."

■ SPECIFICATIONS

- **RATINGS**
- Pilot duty type (NEMA A600/B600)

Circuit	Rated voltage	Current (A)		Voltamperes (VA)	
		Make	Break	Make	Break
SPDT DOUBLE BREAK	120 VAC	60	6	7,200	720
	240 VAC	30	3		
	480 VAC	15	1.5		
	600 VAC	12	1.2		
DPDT DOUBLE BREAK	120 VAC	30	3	3,600	360
	240 VAC	15	1.5		
	480 VAC	7.5	0.75		
	600 VAC	6	0.6		

- **Electronic duty type**
This type is ideal for use in the load range (Zones 1 through 3) shown below.

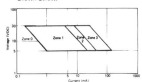

Approved by Standards
UL (File No. E68575)
(Control No. Ind. Cont. Eq. 298Z)
Rating NEMA A600
 600V AC MAX PILOT DUTY*
CSA (File No. LR45746)
Rating NEMA A600
 600V AC MAX PILOT DUTY*

* PILOT DUTY is one of the rating classifications specified for "UL917 clock-operated switches" and refers to large inductive loads with a power factor of 0.35 or less.

- **CHARACTERISTICS**

Operating speed	.04" to 6.56'/sec (at side rotary type)
Operating frequency	Mechanically: 300 operations/min. Electrically: 30 operations/min.
Contact resistance	25mΩ max. (initial)
Insulation resistance	100MΩ min. (at 500 VDC) between each terminal and non-current-carrying part and ground
Temperature rising	30 degree max.
Dielectric strength	1,000 VAC, 50/60Hz for 1 minute between non-continuous terminals; 2,200 VAC, 50/60Hz for 1 minute between each terminal and non-current-carrying metal part and between each terminal and ground
Vibration	Malfunction durability: 10 to 55Hz, .06" double amplitude
Shock	Mechanical durability: Approx. 100G's Malfunction durability: Side rotary type: Approx. 60G's Non side rotary type: Approx. 30G's
Ambient temperature	Operating: D4A-11□□/-12□□ series: +14 to +248° F (D4A-1103/-1203: +23 to +248° F) D4A-13□□/-14□□ series: +14 to +176° F D4A-25□□/-26□□ series: +14 to +194° F D4A-□□□□-T: −40 to 158° F
Humidity	95% RH max.
Degree of protection	NEMA: Types 1, 2, 3, 3R, 4, 4X, 5, 6, 12, 13
	IEC: IP-67
	JIS: Immersion-proof type
Service life	Mechanical: 50,000,000 operations min.
Weight	Approx. 20.6 oz. (D4A-1101)

Model D4A — OMRON
Cat. No. C03-E3-2

- **CONTACT CONFIGURATION**

SPDT DOUBLE BREAK (Must be same polarity)

Without lamp / With lamp

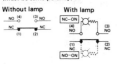

DPDT DOUBLE BREAK (Same polarity each pole)

Sequential operating type:
Pole 1 operates first
Pole 2 operates second
Either CW or CCW or Both

Center neutral operating type:
Pole 1 operates CW
Pole 2 operates CCW

- **OPERATING CHARACTERISTICS**

Type	D4A-☐☐01	D4A-☐☐02	D4A-☐☐03	D4A-☐☐04	D4A-☐☐05	D4A-☐☐06	D4A-☐☐07	D4A-☐☐08	D4A-☐☐09
OF oz. max.	3.5 lb-in	3.5 lb-in	1.7 lb-in	1.7 lb-in	3.5 lb-in	95.2	95.2	95.2	63.5
RF oz. min.	.4 lb-in	.4 lb-in	.3 lb-in	.3 lb-in	—	31.7	31.7	31.7	17.6
PT in. max.	15°	7°	15°	7°	65°	.094	.094	.094	.063
OT in. min.	60°	68°	60°	68°	20°	.201	.201	.201	.201
MD in. max.	5° (7°)	4° (6°)	5° (7°)	4° (6°)	35° (35°)	.023 (.039)	.023 (.039)	.023 (.039)	.016 (.039)
OP in.	—	—	—	—	1.339±.031	1.732±.031	1.614 to 1.870	1.811±.031

Type	D4A-☐☐10	D4A-☐☐11	D4A-☐☐12	D4A-☐☐14	D4A-☐☐15	D4A-☐☐16	D4A-☐☐17	D4A-☐☐18
OF oz. max.	63.5	63.5	3.5	5.3	5.3	5.3	3.5 lb-in	3.5 lb-in
RF oz. min.	17.6	17.6	—	—	—	—	.2 lb-in	.2 lb-in
PT in. max.	.063	.063	3.937 Radius	1.378 Radius	1.969 Radius	1.969 Radius	1st Pole 10° 2nd Pole 18°	15°
OT in. min.	.201	.201	—	—	—	—	57°	60°
MD in. max.	.016 (.039)	.016 (.039)	—	—	—	—	Each Pole 5°	5°
OP in.	2.205±.031	2.185 to 2.441	—	—	—	—	—	—

NOTE: Figure in parenthesis applies to the DPDT double break circuit type (D4A-2☐☐☐).

■ DIMENSIONS (Unit: inch)

NOTES: 1. Unless otherwise specified, a tolerance of ±.016" applies to all dimensions.
2. All dimensions shown here are for reference only.

SWITCH UNITS

- D4A-☐☐01 ~ D4A-☐☐05
  D4A-☐☐17 ~ D4A-☐☐18

- D4A-☐☐06

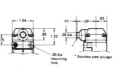
* Stainless steel plunger

- D4A-☐☐07-V

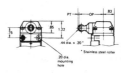
* Stainless steel roller

- D4A-☐☐07-H

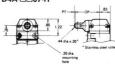
* Stainless steel roller

- D4A-☐☐08

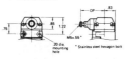
* Stainless steel hexagon bolt

- D4A-☐☐09

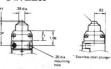
* Stainless steel plunger

- D4A-☐☐10

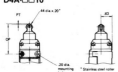
* Stainless steel roller

- D4A-☐☐11

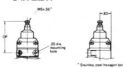
* Stainless steel hexagon bolt

- D4A-☐☐12

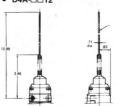
* Stainless steel cat whisker

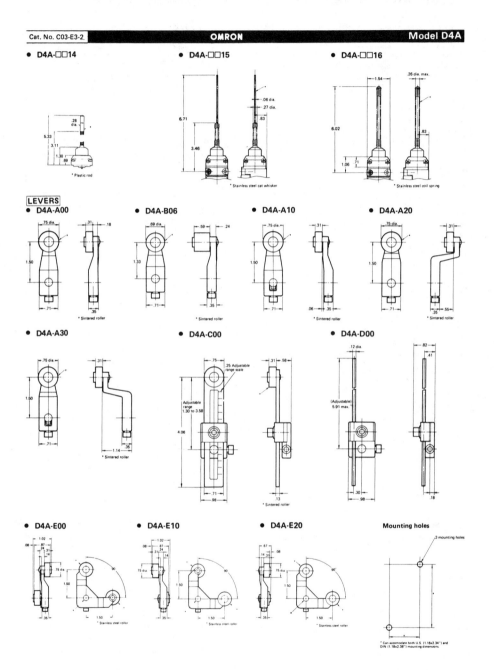

Model D4A — OMRON — Cat. No. C03-E3-2

■ CAM TRACKING

The standard levers and offset levers permit a wide range of cam tracking possibilities as shown below.

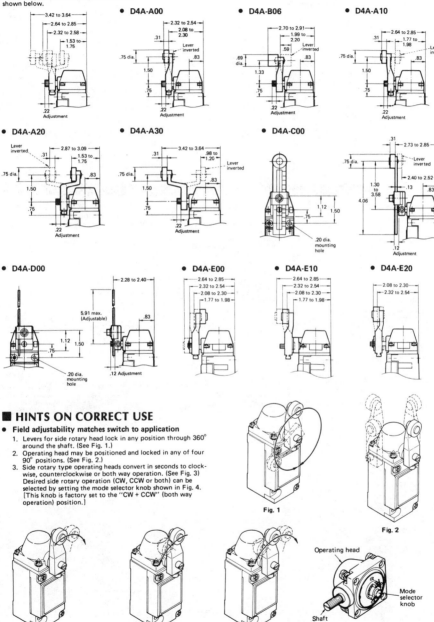

- D4A-A00
- D4A-B06
- D4A-A10
- D4A-A20
- D4A-A30
- D4A-C00
- D4A-D00
- D4A-E00
- D4A-E10
- D4A-E20

■ HINTS ON CORRECT USE

- **Field adjustability matches switch to application**
  1. Levers for side rotary head lock in any position through 360° around the shaft. (See Fig. 1.)
  2. Operating head may be positioned and locked in any of four 90° positions. (See Fig. 2.)
  3. Side rotary type operating heads convert in seconds to clockwise, counterclockwise or both way operation. (See Fig. 3) Desired side rotary operation (CW, CCW or both) can be selected by setting the mode selector knob shown in Fig. 4. [This knob is factory set to the "CW + CCW" (both way operation) position.]

Fig. 1

Fig. 2

Fig. 3 — CW, CCW, CW+CCW

Fig. 4 — Operating head, Mode selector knob, Shaft

Model D4A

● Roller position can be selected

D4A-E00   D4A-E10

D4A-E20

Applicable actuators:
Folk lever lock (D4A-E□□)

● Switches with Viton® seals

Model D4A limit switches are for use in applications where the environment includes fire-resistant synthetic fluids. In addition to almost all fluids, the Viton® sealed type may be used with such industrial fluids as Cellulube, Fyrquell, Houghto-Safe, Pydraul and other special cutting and hydraulic oils.

● Lighting Mode Selection of Indicator Lamp

The lighting mode of the operation indicator lamp incorporated in the switch body can be changed easily between the mode to light when the switch is in the operate state and the mode to light when the switch is in the non-operate state.

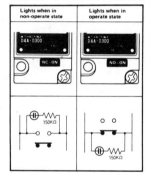

Change the lighting mode as follows:

Gently push the claw securing the lamp section to the right.

Remove the lamp section.

Mount the lamp section so that legend "NC–ON" or "NO–ON" will appear in the display window.

Model D4A — OMRON
Cat. No. C03-E3-2

■ REPLACEMENT PARTS

Switch bodies, receptacles, operating heads and levers may be ordered separately as replacement parts by using the part numbers listed below.

- Side rotary head type

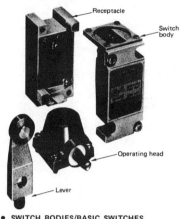

- Side plunger type

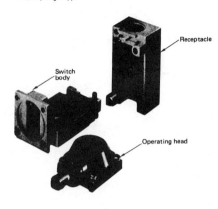

● SWITCH BODIES/BASIC SWITCHES

Circuit	SPDT DOUBLE BREAK				DPDT DOUBLE BREAK		Sequential operating		Center neutral operating	
Indicator lamp	Without Lamp		With Lamp		Without Lamp		Without Lamp		Without Lamp	
Contact rating	Pilot duty	Electronic duty	Pilot duty	Electronic duty	Pilot duty	Electronic duty	Pilot duty	Electronic duty	Pilot duty	Electronic duty
Switch body type	D4A-0100	D4A-0200	D4A-0300	D4A-0400	D4A-0500	D4A-0600	D4A-0700	D4A-0800	D4A-0900	D4A-0000
Basic switch type	D4A-S10	D4A-S20	—	—	D4A-S50	D4A-S60	D4A-S70	D4A-S80	D4A-S90	D4A-S00

● RECEPTACLES

Circuit	SPDT DOUBLE BREAK	DPDT DOUBLE BREAK/ Sequential operating/ Center neutral operating
Type	D4A-1000	D4A-2000

● OPERATING HEADS

	Standard	D4A-0001
Side rotary type	High precision	D4A-0002
	Low torque	D4A-0003
	High precision/Low torque	D4A-0004
	Maintained	D4A-0005
	Sequential operating	D4A-0017
	Center neutral operating	D4A-0018
Side plunger type	Plain	D4A-0006
	Vertical roller	D4A-0007-V
	Horizontal roller	D4A-0007-H
	Adjustable	D4A-0008
Top plunger type	Plain	D4A-0009
	Roller	D4A-0010
	Adjustable	D4A-0011
Wobble lever type	Spring wire	D4A-0012
	Plastic rod	D4A-0014
	Cat whisker	D4A-0015
	Coil spring	D4A-0016

● LEVERS FOR SIDE ROTARY HEAD

Standard-roller front mounted	.75 dia./.31 width	D4A-A00
	.68 dia./.59 width	D4A-B06
Standard-roller back mounted		D4A-A10
Offset-roller front mounted		D4A-A20
Offset-roller back mounted		D4A-A30
Adjustable-roller		D4A-C00
Rod-adjustable		D4A-D00
Fork-rollers on opposite side		D4A-E00
Fork-rollers on opposite side		D4A-E10
Fork-rollers on same side		D4A-E20

NOTE: ALL DIMENSIONS SHOWN IN THIS CATALOG ARE IN UNITS OF INCHES. To convert inches into millimeters multiply by 25.4. To convert ounces into grams multiply by 28.353.

APPENDIX B

TECHNICAL INFORMATION

GENERAL CHARACTERISTICS OF PROXIMITY SENSOR

■ How to Express Sensing Distance

In general, proximity sensors have a three dimensional range in the vicinity of their respective sensing surfaces, and the sensing distance for each type of proximity sensor is defined as follows according to the reference position for measurement and the approaching direction of the object to be sensed.

Type of sensor	Illustration	Explanation
Column head & square pillar head types	**Vertical sensing distance**	The sensing distance is expressed by the distance measured from the reference plane by bringing the object to be sensed closer in the direction of the reference axis (perpendicularly to the sensing surface).
Column head & square pillar head types	**Horizontal sensing distance**	The sensing distance is expressed by the distance measured from the reference axis by moving the object to be sensed parallel to the reference plane (i.e., sensing surface). This distance can be expressed as the locus of operating point, since it varies with the passing position of the object (distance from the reference plane).
Through head type		With the through head and other similar types (TL-F20), the reference plane for measurement is determined as shown on the left, and then the sensing distance is measured by bringing the rod-shaped object to be sensed close in the direction of the reference axis.
Grooved head type		With the grooved head type, an object to be sensed such as sheet metal is in most cases passed through the groove of the sensing head for sensing. For this reason, the distance of the object to be sensed inserted into the groove is measured from the reference plane as shown on the left. X': Resetting distance. X: Sensing distance.

■ Salient Characteristics

General characteristics of proximity sensors are shown below with the high-frequency oscillation type (front detection type) proximity sensor taken as an example.

Item	Illustration	Characteristics
Size of object to be sensed vs. sensing distance	With the thickness of a square sheet iron being constant (t=1mm), the sensing distance Xmm is measured by changing the length of one side of the object. (Front sensing type) Ex.) TL-N	With objects larger than the standard object to be sensed, the sensing distance is almost constant. With the through head type, the sensing distance is shown by the data affected by the diameter and length of the object to be sensed. (i.e., columnar metal rod).
Thickness of object to be sensed vs. sensing distance	The sensing distance Xmm is measured by changing the thickness of the specified standard object to be sensed. (Front sensing type) Object: Aluminum, Sensing distance: 5mm	With the high-frequency oscillation type, when the thickness of a nonferrous metal is about 0.01mm, the sensing distance with the nonferrous metal becomes much the same as that with a ferrous metal. With ferrous metals (iron, etc.), no change will occur in the sensing distance so long as the thickness of the metal exceeds 1mm.
Influences of object to be sensed material and type of plating on sensing distance	Influences of the shape, size and material of the standard object to be sensed and type of plating on the sensing distance Xmm are measured. Material / Sensing distance: Iron 100%, Stainless steel Approx. 50%, Brass Approx. 40%, Aluminum Approx. 30%, Copper Approx. 28%	Influences of material other than iron and the type of plating on the sensing distance differ from one type of sensor to another. The versions capable of sensing all kinds of metals are less affected by plating.

PRECAUTIONS REQUIRED IN PROPER SENSOR SELECTION

DESIGN OF OBJECT TO BE SENSED ((EX.) WITH TYPE TL-N)

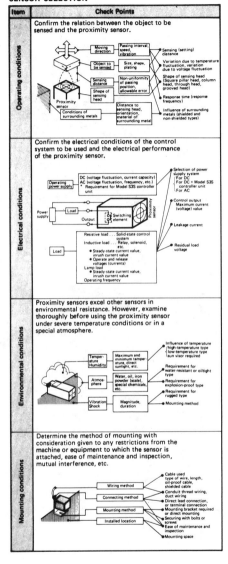

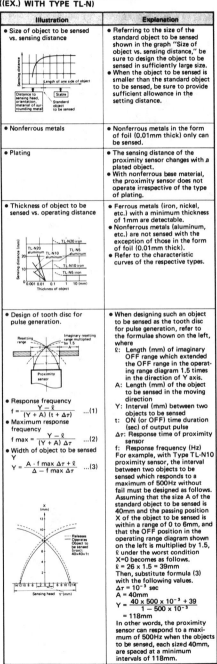

GLOSSARY

Term	Definition
Sensing distance	"Sensing distance" refers to the distance at which the proximity sensor operates (or releases) when measured from the reference position (or reference plane) by moving the object to be sensed in the specified manner. The item "Sensing distance" under "SPECIFICATIONS" indicates the value(s) when measured with the standard object to be sensed.
Setting distance	"Setting distance" refers to the distance from the sensing surface to the passing position of the object to be sensed which permits the proximity sensor to operate without any malfunctions due to temperature or voltage fluctuation. The item "Setting distance" under "SPECIFICATIONS" indicates the value(s) when measured with the standard object to be sensed.
Standard object to be sensed	"Standard object to be sensed" refers to the object to be sensed of specified shape, size and material which is used as the standard to examine the fundamental performance of the proximity sensor.
Differential distance (Hysteresis)	"Differential distance (Hysteresis)" refers to the absolute value of a difference between the sensing and resetting distances. The item "Differential distance (Hysteresis)" under "SPECIFICATIONS" indicates the value(s) when measured with the standard object to be sensed.
Response time	"Response time" refers to the time from the placement of the proximity sensor in the operable state subsequent to the entry of the object to be sensed within the operating range of the sensor until the appearance of the output signal (i.e., time t_1) and the time from the departure of the object to be sensed from the operating range until the disappearance of the output signal (i.e., time t_2).
Response frequency	"Response frequency" refers to the frequency of outputs by the proximity sensor per second in response to the movement of each object to be sensed when brought closer to the sensor. The method of measurement is as outlined on the left.
Leakage current	"Leakage current" refers to the current value when measured with the output stage switching element in the OFF state.
Variation due to temperature fluctuation	"Variation due to temperature fluctuation" refers to a change in sensing distance when the ambient operating temperature changes within a permissible range and is expressed as the rate of variation (in percentage) with the sensing distance at 20°C taken as 100%. $\pm \frac{B}{A} \times 100\ (\%)$
Variation due to voltage fluctuation	"Variation due to voltage fluctuation" refers to a change in sensing distance when the supply voltage changes within a permissible range and is expressed as the rate of variation (in percentage) with the sensing distance at the rated voltage taken as 100%. $\pm \frac{B}{A} \times 100\ (\%)$
Operating current	The maximum value of current to be consumed by the proximity sensor alone when it is not sensing any object.

Compact Type

OMRON PROXIMITY SENSOR

Cat. No. **D01**-E3-3

Model **TL-X**

Cylindrical Type Proximity Sensor Boasting Ultra Small Size and High Performance

■ FEATURES
- Wide operating voltage and temperature range.
- AC switching types have high dielectric strength.
- All DC switching types are equipped with short-circuit and reverse polarity protection.

■ AVAILABLE TYPES

(TL-□□□□: Add the appropriate code when placing your order, e.g., TL-X1E1)

Type			Shielded type				Non-shielded type		
		O.D.	M8	M12	M18	M30	M12	M18	M30
Output form**	Sensing distance		1mm	2mm	5mm	10mm	5mm	10mm	18mm
DC switching type	NPN	N.O.	-X1E1	-X2E1	-X5E1*	-X10E1*	-X5ME1*	-X10ME1*	-X18ME1*
		N.C.	-X1E2	-X2E2	-X5E2*	-X10E2*	-X5ME2*	-X10ME2*	-X18ME2*
	PNP	N.O.	-X1F1	-X2F1	-X5F1*	-X10F1*	-X5MF1*	-X10MF1*	-X18MF1*
		N.C.	-X1F2	-X2F2	-X5F2*	-X10F2*	-X5MF2*	-X10MF2*	-X18MF2*
AC switching type		N.O.	—	-X2Y1	-X5Y1*	-X10Y1*	-X5MY1*	-X10MY1*	-X18MY1*
		N.C.	—	-X2Y2	-X5Y2*	-X10Y2*	-X5MY2*	-X10MY2*	-X18MY2*

NOTES:
* The types identified with an asterisk are available with different frequencies, when using two sensors adjacent to each other. (Add suffix "5" to the part number when placing your order.)
** Refer to "OUTPUT STAGE CIRCUIT DIAGRAMS."
N.O.: Normally Open
N.C.: Normally Closed

――――― OMRON ―――――

■ SPECIFICATIONS
● RATINGS

Type										
	DC switching type	NPN	TL-X1E1	TL-X2E1	TL-X5E1	TL-X10E1	TL-X5ME1	TL-X10ME1	TL-X18ME1
			TL-X1E2	TL-X2E2	TL-X5E2	TL-X10E2	TL-X5ME2	TL-X10ME2	TL-X18ME2
		PNP	TL-X1F1	TL-X2F1	TL-X5F1	TL-X10F1	TL-X5MF1	TL-X10MF1	TL-X18MF1
			TL-X1F2	TL-X2F2	TL-X5F2	TL-X10F2	TL-X5MF2	TL-X10MF2	TL-X18MF2
	AC switching type		—	TL-X2Y1	TL-X5Y1	TL-X10Y1	TL-X5MY1	TL-X10MY1	TL-X18MY1
Item			—	TL-X2Y2	TL-X5Y2	TL-X10Y2	TL-X5MY2	TL-X10MY2	TL-X18MY2
Rated voltage	DC switching type		10 to 40 VDC*, **						
	AC switching type		45 to 260 VAC						
Operating current	DC switching type		10mA max.						
	AC switching type		—	2mA max.					
Materials sensed			All metals (refer to "Characteristic Data.")						
Sensing distance			1mm	2mm	5mm	10mm	5mm	10mm	18mm
Setting distance (with standard object)			0 to 0.8mm (Iron: 8x8x1t)	0 to 1.6mm (Iron: 12x12x1t)	0 to 4.0mm (Iron: 18x18x1t)	0 to 8.0mm (Iron: 30x30x1t)	0 to 4.0mm (Iron: 15x15x1t)	0 to 8.0mm (Iron: 30x30x1t)	0 to 14.0mm (Iron: 54x54x1t)
Differential distance (Hysteresis)			10% max. of sensing distance						
Response frequency***	DC switching type		1kHz min.	800Hz min.	350Hz min.	250Hz min.	400Hz min.	200Hz min.	100Hz min.
	AC switching type		20Hz min.						
Control output** (Switching capacity)	DC switching type		200mA max.						
	AC switching type		5 to 200mA						

NOTES:
* For all DC switching types except TL-X1□□, unsmoothed full-wave rectified power source of 24 VDC ±20% (mean value) may be used.
** Refer to "Operating Voltage vs. Temperature Characteristic" and "Load Current vs. Temperature Characteristics."
*** With the DC switching type, the response frequency values shown are mean values.

Approved by Standard
UL (DC switching types: File No. E61312
  AC switching types: File No. E76675)

Model TL-X   OMRON   Cat. No. D01-E3-3

● CHARACTERISTICS

Type \ Item	TL-X1□□	TL-X2□□	TL-X5□□	TL-X10□□	TL-X5M□□	TL-X10M□□	TL-X18M□□
Variation due to temperature fluctuation	±15% max. of sensing distance at 20°C within a temperature range of −40 to +85°C ±10% max. of sensing distance at 20°C within a temperature range of −25 to +70°C							
Variation due to voltage fluctuation	DC switching type: ±2.5% max. of sensing distance at 12/24 VDC when operated within 10 to 40 VDC AC switching type: ±1% max. of sensing distance at 100/200 VAC when operated within 45 to 260 VAC						
Operating current (Leakage current)	Refer to "Leakage current characteristics" (AC switching type only)						
Residual voltage	Refer to "Residual Load Voltage Characteristics".						
Insulation resistance	50MΩ min. (at 500 VDC)						
Dielectric strength	DC switching type: 1,000 VAC, 50/60Hz for 1 minute AC switching type: 4,000 VAC, 50/60Hz for 1 minute						
Vibration	Mechanical durability: 10 to 55Hz, 1.5mm double amplitude						
Shock	Mechanical durability: 500m/s² (approx. 50G's)	Mechanical durability: 1,000m/s² (approx. 100G's)		Mechanical durability: 500m/s² (approx. 50G's)	Mechanical durability: 1,000m/s² (approx. 100G's)		
Degree of protection	IP67 (IEC 144), NEMA types 1, 4, 6, 12, 13						
Ambient operating temperature	−40 to +85°C						
Humidity	35 to 95% RH						
Weight*	Approx. 45g	Approx. 125g	Approx. 165g	Approx. 310g	Approx. 120g	Approx. 160g	Approx. 300g

* The weight includes a 2m cable attached as an accessory.

● OUTPUT STAGE CIRCUIT DIAGRAMS

DC switching type
NPN (Sink type)

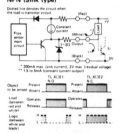

* 200mA max. (sink current), 2V max. (residual voltage)
** 1.5 to 5mA (constant current output)

PNP (Source type)

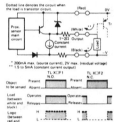

* 200mA max. (source current), 2V max. (residual voltage)
** 1.5 to 5mA (constant current output)

AC switching type

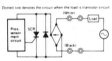

● CHARACTERISTIC DATA (Typical examples)

Operating Range Diagram
TL-X1□□, TL-X2□□

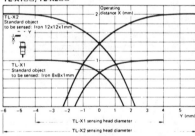

TL-X5□□, TL-X10□□

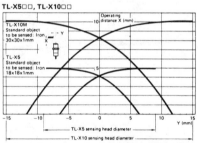

TL-X5M□□, TL-X10M□□, TL-X18M□□

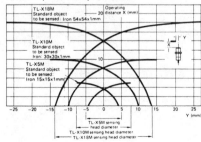

Operating Voltage vs. Temperature Characteristics
TL-X□E(F)

Load Current vs. Temperature Characteristics
TL-X□Y

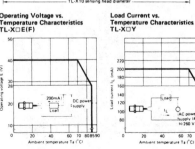
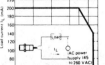

Residual Load Voltage Characteristics
100 VAC   **200 VAC**

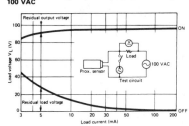 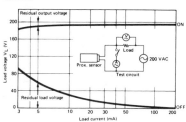

NOTE: When the current rating of the load is less than 5mA, false operating may occur. This is normal, and the problem can be cured by installing a bleeder resistor in parallel with the load. Use the formulas given here to calculate the power rating and value of the resistor.

$$R \leq \frac{V_s}{5-i} \ (k\Omega)$$
$$P > \frac{V_s^2}{R} \ (mW)$$

P : Power rating of bleeder resistor
i : Load current (mA)
V_s : Supply voltage (V)

Operating Current (Leakage Current) Characteristics*

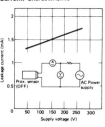

NOTE: * When an AC sensor is in the OFF state, a very small amount of current flows to operate the internal circuit of the sensor.

Sensing Distance vs. Size and Material of Object
TL-X1

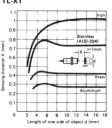

Sensing Distance vs. Size and Material of Object

TL-X2   TL-X5   TL-X10

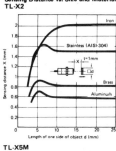   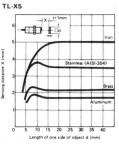   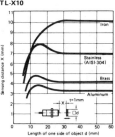

TL-X5M   TL-X10M   TL-X18M

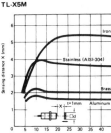   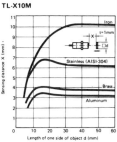 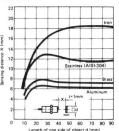

NOTE: If the object to be sensed is nonferrous metal, the sensing distance of the proximity sensor decreases. However, with a piece of foil measuring about 0.01mm in thickness, the sensing distance is equivalent to that with ferrous metals. Note that the proximity sensor cannot sense extremely thin evaporated films and non-conductive objects.

Model TL-X — OMRON

Cat. No. D01-E3-3

■ DIMENSIONS (Unit: mm)

- **TL-X1□□**

- **TL-X2E(F)□**
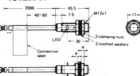

- **TL-X2Y□**
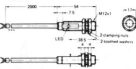

- **TL-X5ME(F)□**
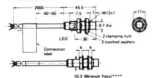

- **TL-X5MY□**
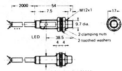

- **TL-X5E(F)□**
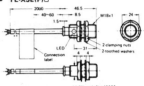

- **TL-X5Y□**
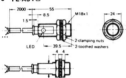

- **TL-X10ME(F)□**
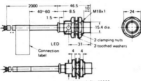

- **TL-X10MY□**
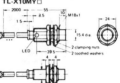

- **TL-X10E(F)□**
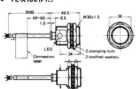

- **TL-X10Y□**
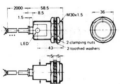

- **TL-X18ME(F)□**
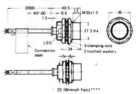

- **TL-X18MY□**
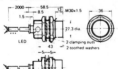

- **Mounting fixture**
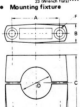

NOTES:
- * Polyvinyl chloride insulated cable: 3-0.14mm², Outer diameter: 3.5mm, Standard length: 2m, Extendable cable length: 200m max. (independently through metal conduit)
- ** Polyvinyl chloride insulated cable: 3-0.5mm², Outer diameter: 6mm, Standard length: 2m, Extendable cable length: 200m max. (independently through metal conduit)
- *** Polyvinyl chloride insulated cable: 2-0.5mm², Outer diameter: 6mm, Standard length: 2m, Extendable cable length: 200m max. (independently through metal conduit)
- **** The part of the sensor body where the LED indicator is located and its opposite side are flatly cut to facilitate wrenching when mounting or adjusting the sensor.

Mounting holes

TL-X1□□	TL-X2□□ / TL-X5M□□	TL-X5□□ / TL-X10M□□	TL-X10□□ / TL-X18M□□
9 $^{+0.5}_{-0}$ dia.	13 $^{+1}_{-0}$ dia.	19 $^{+1}_{-0}$ dia.	31 $^{+1}_{-0}$ dia.

Dimensions Type	A	B	C	D (dia.)	E	F (Hexagon bolt)	Applicable types
Y92E-B8	18±0.2	10 max.	18	8	28 max.	M4×20	TL-X1E(F)
Y92E-B12	24±0.2	12.5 max.	20	12	37 max.	M4×25	TL-X2E(F)(Y), TL-X5ME(F)(Y)
Y92E-B18	32±0.2	17 max.	30	18	47 max.	M5×32	TL-X5E(F)(Y), TL-X10ME(F)(Y)
Y92E-B30	45±0.2	17 max.	50	30	60 max.	M5×50	TL-X10E(F)(Y), TL-X18ME(F)(Y)

Model TL-X

■ CONNECTIONS

● DC switching type (NPN)
Single unit control

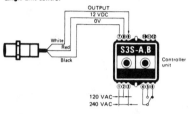

2-unit control

● AC switching type

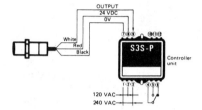

NOTE: See the catalog for Model S3S Controller Unit. (Model S3S is UL recognized.)

NOTE: For each proximity sensor, use Type TL-X□E2 only.

■ HINTS ON CORRECT USE

● Influence of surrounding metals

When mounting a proximity sensor flush with a metallic panel, be sure to provide a minimum distance as shown in Table 1 below, to prevent the sensor from being affected by metal objects other than the object to be sensed.

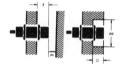

● Influence of plating (reference value)
% against sensing distance with unplated object

Type of plating	Base material Iron	Brass
Unplated	100	100
Zn 5~15μ	90~120	95~105
Cd 5~15μ	100~110	95~105
Ag 5~15μ	60~90	85~100
Cu 10~20μ	70~95	95~105
Cu 3~10μ	—	95~105
Cu (5~10μ) + Ni (10~20μ)	75~95	—
Cu (5~10μ) + Ni (10μ) + Cr (0.3μ)	75~95	—

Table 1. Effects of surrounding metals (Unit: mm)

Item \ Type	TL-X10□	TL-X20□	TL-X50□	TL-X100□	TL-X5M□□	TL-X10M□□	TL-X18M□□
ℓ	0	0	0	0	15	22	30
d	8	12	18	30	40	55	90
D	0	0	0	0	15	22	30
m	4	8	20	40	20	40	70

● Mutual interference

Be sure to space the two sensors at a distance greater than that shown in Table 2, to prevent mutual interference.

Table 2. Mutual interference (Unit: mm)

Item \ Type	TL-X10□	TL-X20□	TL-X50□	TL-X100□	TL-X5M□□	TL-X10M□□	TL-X18M□□
A	20	30	50(30)	100(50)	120(60)	200(100)	300(100)
B	15	20	35(18)	70(35)	100(50)	110(60)	200(70)

NOTE: The values in parentheses apply to the alternate frequency ("5" suffix) type. Refer to "AVAILABLE TYPES."

● Mounting

Do not tighten the proximity sensor excessively.

NOTE:
Tightening strength differs depending on the location of the clamping nut from the face of the sensing head. The table on the right shows the tightening torque applicable to each type when the nut is to be tightened in location A (e.g., within 11mm from the face of the sensing head) or in location B as shown in the drawing shown above.

(Shielded type)

(Non-shielded type)

Strength Type	A Dimensions	A Torque	B Torque
TL-X1E(F)□	11mm	20 kg-cm	30 kg-cm
TL-X2E(F/Y)□	17mm	60 kg-cm	100 kg-cm
TL-X5E(F/Y)□	21mm	150 kg-cm	500 kg-cm
TL-X10E(F/Y)□	24mm	400 kg-cm	1,500 kg-cm
TL-X5ME(F/Y)□	10mm	60 kg-cm	100 kg-cm
TL-X10ME(F/Y)□	11mm	150 kg-cm	500 kg-cm
TL-X18ME(F/Y)□	11mm	400 kg-cm	1,500 kg-cm

GLOSSARY

Term	Definition
Sensing distance	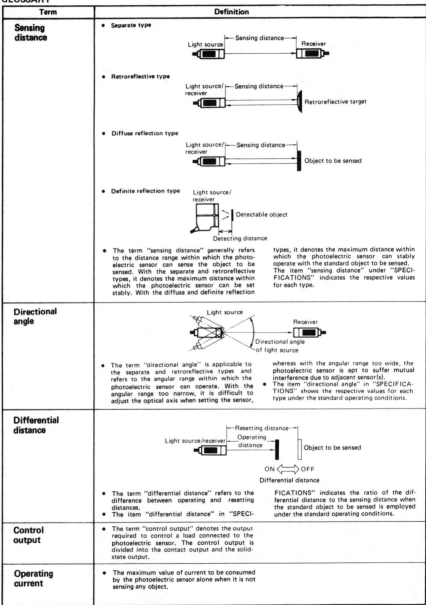 • Separate type • Retroreflective type • Diffuse reflection type • Definite reflection type • The term "sensing distance" generally refers to the distance range within which the photoelectric sensor can sense the object to be sensed. With the separate and retroreflective types, it denotes the maximum distance within which the photoelectric sensor can be set stably. With the diffuse and definite reflection types, it denotes the maximum distance within which the photoelectric sensor can stably operate with the standard object to be sensed. The item "sensing distance" under "SPECIFICATIONS" indicates the respective values for each type.
Directional angle	• The term "directional angle" is applicable to the separate and retroreflective types and refers to the angular range within which the photoelectric sensor can operate. With the angular range too narrow, it is difficult to adjust the optical axis when setting the sensor, whereas with the angular range too wide, the photoelectric sensor is apt to suffer mutual interference due to adjacent sensor(s). • The item "directional angle" in "SPECIFICATIONS" shows the respective values for each type under the standard operating conditions.
Differential distance	• The term "differential distance" refers to the difference between operating and resetting distances. • The item "differential distance" in "SPECIFICATIONS" indicates the ratio of the differential distance to the sensing distance when the standard object to be sensed is employed under the standard operating conditions.
Control output	• The term "control output" denotes the output required to control a load connected to the photoelectric sensor. The control output is divided into the contact output and the solid-state output.
Operating current	• The maximum value of current to be consumed by the photoelectric sensor alone when it is not sensing any object.

Model TL-X — OMRON

General Information

- **Connections for AC-type proximity sensors**
  (1) Two proximity sensors may be wired in series. This will increase the voltage across the sensors, and decrease the voltage available to the load. Some loads may fail to operate in this mode.
  (2) Proximity sensors may be connected in parallel. The leakage current through the load will be the sum of the leakage current for each sensor. This may cause false operation of the load. If two or more proximity sensors are wired in parallel and are in the on-state, the load may turn off for about 150msec when one of the sensors turns off. Therefore, we do not recommend this arrangement.
  (3) When proximity sensors are wired in series or parallel with contact-type switches, problems will occur. System response time may slow down, and the proximity sensors may not have enough voltage to operate properly. We do not recommend using proximity sensors in this manner.

- **Use of metallic conduit**
  If a high voltage or power line runs near the proximity sensor cable, be sure to wire the sensor cable through a metallic conduit to prevent the sensor from malfunctioning or damage.

- **Surge protection**
  The proximity sensor is provided with a surge suppressor circuit. However, if any large surge generating source (e.g., motor, welding machine) exists in the vicinity of the proximity sensor, it is recommended to insert such a surge suppressor as a varistor, into the surge generating source.

NOTE: ALL DIMENSIONS SHOWN IN THIS CATALOG ARE IN UNITS OF MILLIMETERS.
To convert millimeters into inches multiply by 0.03937. To convert grams into ounces multiply by 0.03527.

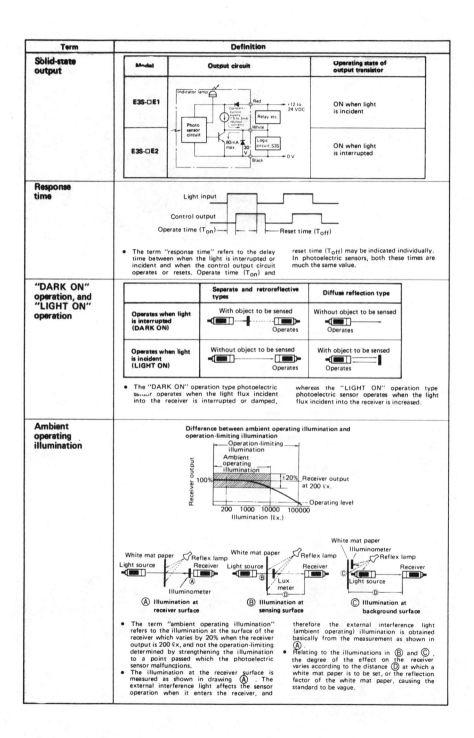

OMRON PHOTOELECTRIC SENSOR
Cat. No. E04-E3-2
Model **E3S-G**

Grooved Head Type Ideal for Mark Sensing and Positioning

■ FEATURES
- Stable sensing of marks on transparent sheets (E3S-GM5).
- Ideal for positioning of opaque and translucent objects and label sensing (E3S-GS3).
- High-speed response time of 2msec max.
- Rugged, highly sealed die-cast case.
- Directly switches 80mA max.

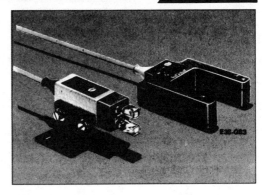

■ AVAILABLE TYPES

Type / Operating stage of output transistor	Mark sensing	General use
Sensing distance (Width of groove)	5mm	3cm
ON when light is incident	E3S-GM5E1	E3S-GS3E1
ON when light is interrupted	E3S-GM5E2	E3S-GS3E2

■ SPECIFICATIONS

● RATINGS

Item / Type	E3S-GM5E□	E3S-GS3E□
Rated voltage	12 VDC−10% to 24 VDC+10%, Ripple (p-p): 10% max.	
Operating current	40mA max.	
Sensing distance	5mm	3cm
Object to be sensed	Marks on transparent sheet* (2x3mm min.)	Opaque and translucent materials (3mm dia. min.)
Control output	Output (source) current: 1.5 to 3mA Load (sink) current: 80mA (Residual voltage: 1V max.)	
Response time	2msec max.	

NOTE: * The switch performs stable detection if the transmissivity of the transparent sheet is 30% min. and that of the mark is 50% max. or vice versa.

● CHARACTERISTICS

Insulation resistance/Dielectric strength	These data are excluded as the sensor is equipped with a capacitor grounding.
Vibration	Mechanical durability: 10 to 55Hz, 1.5mm double amplitude (in X, Y, Z directions, respectively for 2 hours)
Shock	Mechanical durability: 500m/s² (approx. 50G's) (in X, Y, Z directions, respectively 10 times)
External interference light* — Incand. lamp	3,000 lx. max.
External interference light* — Sunlight	10,000 lx. max.
Ambient temperature	Operating: −25 to +55°C Storage: −40 to +70°C
Humidity	Operating: 45 to 85% RH Storage: 35 to 95% RH
Degree of protection	Water-resistant type (JIS C 0920), IP66 (IEC 144), NEMA types 1, 4, 4X, 12
Weight**	E3S-GM5: Approx. 150g, E3S-GS3: Approx. 140g

NOTES: * External interference light denotes the illumination at the surface of the receiver which varies by 20% when the receiver output is 200 lx. It should not be construed as the operational limit illumination.
** With a standard 2m cable.

● OUTPUT STAGE CIRCUIT DIAGRAM
Output circuit

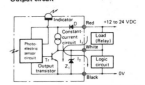

D : Reverse current preventing diode V_R = 30V
Z_1 : Surge protective zener diode V_Z = 30V
I_1 : Constant-current output 1.5 to 3mA
I_2 : Load current 80mA max.
Tr : Output transistor I_1 =80mA max.

Operating state of output transistor	E3S-G□□E1	E3S-G□□E2
Relay, etc.	Operates when light is incident	Operates when light is interrupted
Logic circuit	High when light is interrupted	High when light is incident
Controller units — S3S-A10 / S3S-B10	Output relay operates when light is interrupted	Output relay operates when light is incident
Controller units — S3S-P10	Output relay operates when light is incident	Output relay operates when light is interrupted
Operation indicator	ON when light is incident	ON when light is incident

Approved by Standard
UL (File No. E61312)

Model E3S-G OMRON   Cat. No. E04-E3-2

● CHARACTERISTIC DATA

Passing Speed vs. Mark Width Characteristics
E3S-GM5

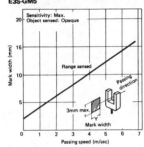

Passing Speed vs. Width of Object Sensed Characteristics
E3S-SG3

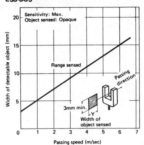

■ DIMENSIONS (Unit: mm)

● E3S-GM5E1, E3S-GM5E2

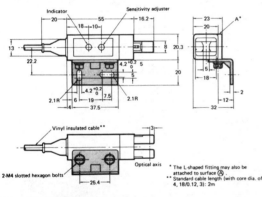

● E3S-GS3E1, E3S-GS3E2

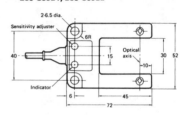

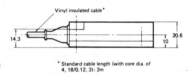

* The L-shaped fitting may also be attached to surface Ⓐ.
** Standard cable length (with core dia. of 4, 18/0.12, 3): 2m

* Standard cable length (with core dia. of 4, 18/0.12, 3): 2m

■ CONNECTIONS

● **When using Model S3S controller unit**
NOTE: See the catalog for Model S3S Controller Unit. (Model S3S is UL recognized.)

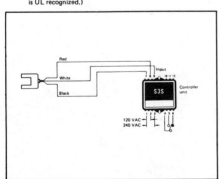

● **When switching loads directly**

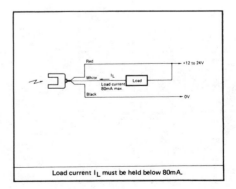

Load current I_L must be held below 80mA.

219

- **Logic circuit interface**

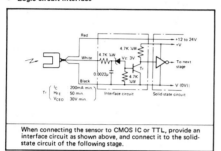

When connecting the sensor to CMOS IC or TTL, provide an interface circuit as shown above, and connect it to the solid-state circuit of the following stage.

■ HINTS ON CORRECT USE

- Adjustment method

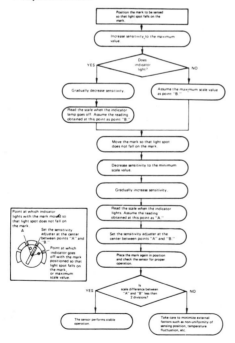

- Other precautions
1. Note that collection of water, dust etc. on the light source/receiver surface may cause the sensor to operate unstably. Periodically wipe dust, water, etc. off the light source/receiver surface.
2. When sensing marks on a transparent sheet with the Type E3S-GM5 photoelectric sensor, make sure that the transparent sheet passes the center of the groove width. If the transparent sheet shakes and is likely to touch the light source/receiver surface, provide the guides as shown in the following drawing to prevent the transparent sheet from shaking.

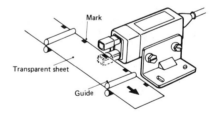

Model E3S-G — OMRON
Cat. No. E04-E3-2

Application Examples

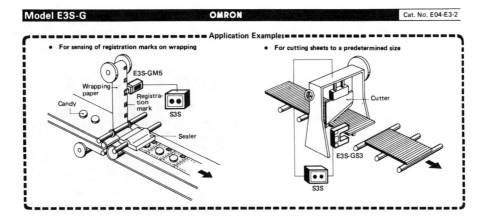

- For sensing of registration marks on wrapping
- For cutting sheets to a predetermined size

NOTE: ALL DIMENSIONS SHOWN IN THIS CATALOG ARE IN UNITS OF MILLIMETERS.
To convert millimeters into inches multiply by 0.03937. To convert grams into ounces multiply by 0.03527.

OMRON CONTROLLER UNIT

Cat. No. Q30-E3-4
Model **S3S**

Miniature Power Supply With Control Functions

■ FEATURES

- Small size with 100mA DC output
- Timer, memory and one-shot delay operations are available in the same unit.
- LED operation and power indicators
- Wide operating temperature and voltage ranges
- Ideal for proximity and photoelectric sensors
- Easy to wire

■ AVAILABLE TYPES

Control function / Output current	Standard	Multi-function		Power
	Standard operation 12 VDC power supply	Timer/memory operation 12 VDC power supply	Timer/one-shot delay operation 12 VDC power supply	Standard operation 24 VDC power supply
		Delay time range (variable)		
		0.1 to 1sec	0.1 to 10sec	
100mA	S3S-A10-US	S3S-B10-US*	S3S-C10	S3S-P10-US

NOTES: 1. All units are supplied with sockets.
2. * 1 to 10sec version is also available, Type S3S-B10-002.

─────── OMRON ───────

■ SPECIFICATIONS

● RATINGS

Item		Type	S3S-A10-US	S3S-B10-US	S3S-C10	S3S-P10-US
AC input	Rated voltage		120/240 VAC, 50/60Hz			
	Operating voltage range		85 to 110% of rated voltage			90 to 110% of rated voltage
	Power consumption		8VA max.			
DC output	Voltage		12 VDC ±10%			24 VDC ±10%
	Maximum current		100mA			
	Output voltage fluctuation due to	Load**	4% max.	5% max.		4% max.
		Supply voltage***	±0.5% max.			±1% max.
		Temperature****	±3% max.			
	Ripple		Refer to "CHARACTERISTIC DATA"			
	Regulation method		Switching			
	Short-circuiting protection		Equipped			
Control output	Capacity*****		250 VAC 5A	250 VAC 3A		250 VAC 5A
	Delay time setting range		—	ON time: 0.1 to 1sec. (variable) OFF time: 0.1 to 1sec. (variable)	1) 1sec./10sec. selector switch incorporated. 2) 1sec. range: 0.1 to 1sec. (variable) 10sec. range: 1 to 10sec. (variable)	—
Signal input	Minimum input time		20msec.	Input 1: 2msec. Input 2: 20msec.	1sec. range: 2ms 10sec. range: 20ms	20msec.
	Signal level		Positive polarity H: 4 to 12V****** L: 0 to 1V (for set and reset of inputs 1 and 2)			Pull-in current: 30mA min. Release current: 1mA max.
	Input impedance		Approx. 4.7kΩ			Approx. 1kΩ

NOTES:
* All data related to characteristics are measured one hour after the application of power to the controller unit.
** Output voltage fluctuation due to load is measured at an ambient temperature of 20°C with the load changed within 0 to 100% one hour after the application of rated voltage to the controller unit.
*** Output voltage fluctuation due to supply voltage is measured at an ambient temperature of 20°C with a 100% load within ±10% change of rated voltage one hour after application of power to the controller unit.
**** Output voltage fluctuation due to temperature is measured at ambient temperatures ranging from −10 to +55°C one hour after the application of rated voltage to the controller unit and is expressed by percentage with the data at 20°C taken as 100.
***** SPDT relay contact.
****** 6 to 12V when the set input (Input 1) is used at an input time of 5msec max.

Approved by standard
UL (File No. E61312)

Model S3S — OMRON — Cat. No. Q30-E3-4

CHARACTERISTICS

Insulation resistance	10MΩ min. (at 500 VDC) between AC input and DC output terminals and between current-carrying and non-current-carrying parts
Dielectric strength	2,000 VAC, 50/60Hz for 1 minute between AC input and DC output terminal and between current-carrying and non-current-carrying parts
Vibration	Mechanical durability: 10 to 30Hz; 1.5mm double amplitude (in X, Y, Z directions, respectively for 2 hours) Malfunction durability: 10 to 30Hz; 1mm double amplitude (in X, Y, Z directions, respectively for 2 hours)
Shock	Mechanical durability: 100m/sec² (approx. 10G's) (in X, Y, Z directions, respectively 5 times) Malfunction durability: 50m/sec² (approx. 5G's) (in X, Y, Z directions, respectively 5 times)
Ambient temperature	Operating: −10 to +55°C
Humidity	45 to 85% RH
Service life (output relay)	Mechanically: 20,000,000 operations min. Electrically: See "CHARACTERISTIC DATA."
Weight	S3S-A10-US: Approx. 300g; S3S-B10-US/S3S-C10/S3S-P10-US: Approx. 320g

CHARACTERISTIC DATA

Electrical service life

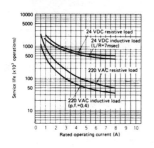

Supply voltage vs. ripple voltage characteristics
S3S-A10-US/-B10-US/-P10-US

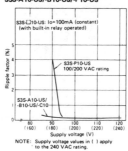

NOTE: Supply voltage values in ( ) apply to the 240 VAC rating.

Output current vs. output voltage characteristics

S3S-A10-US/-B10-US/-C10

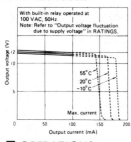

S3S-P10-US

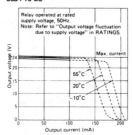

■ OPERATIONS (NOTE: For inputs 1 and 2 in the following diagrams, refer to "CONNECTIONS.")

● Type S3S-A10-US

The standard type controller unit performs two basic functions; one is to convert AC line voltage into a constant voltage of 12 VDC and supply the DC voltage as an output, and the other is to provide relay contact output by driving the built-in relay with an external signal.

(1) When using either input 1 or 2

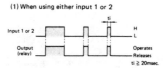

(2) When using both inputs 1 and 2 (AND operation)

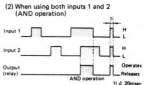

● Type S3S-B10-US

The multi-function type controller unit performs timer (ON-delay, OFF-delay) operation and memory operation in addition to the standard operation.

● Timer operation
For ON-delay or OFF-delay operation, only signal input 1 (terminal ⑨) is used. Signal input 2 (terminal ⑫) is not used. In this case, signal input 1 serves as an input signal for the transistor which drives the built-in relay.

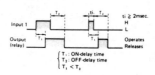

T_1: ON-delay time
T_2: OFF-delay time
$T_1 < T_2$

Cat. No. Q30-E3-4   **OMRON**   Model **S3S**

- **Memory operation**

The controller unit performs memory operation with signal input 1 used as the set input and signal input 2 as the reset input and with the ON/OFF delay variable resistor of the timer set at the "MIN." position. On application of power to the controller unit, the memory circuit is stabilized when the output relay is in the OFF state. On application of signal input 1, the memory circuit is placed in the set (relay ON) state, and on application of signal input 2, the circuit is placed in the reset (relay OFF) state. When both set (input 1) and reset (input 2) signals are input simultaneously, the set input takes precedence over the reset input.

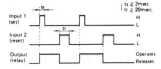

- **Type S3S-C10**

In addition to the standard operation, selection of ON-delay, OFF-delay and one-shot delay operations, as well as input signal inversion are possible by the built-in mode selector switch. The delay time is variable within range of 0.1 to 10sec.

Selection method

- Setting of H-ON/L-ON selector switch

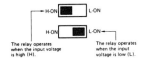

The relay operates when the input voltage is high (H).

The relay operates when the input voltage is low (L).

- Selection of ON-delay, OFF-delay or one-shot delay operation

ON-delay selection   OFF-delay selection   One-shot delay selection

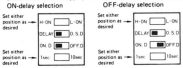

Set either position as desired

Set either position as desired

Irrelevant since both ON-delay and OFF-delay functions are disabled.

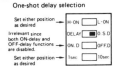

Set either position as desired

- **Setting of delay time**

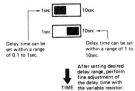

Delay time can be set within a range of 0.1 to 1sec.

Delay time can be set within a range of 1 to 10sec.

After setting desired delay range, perform fine adjustment of the delay time with the variable resistor.

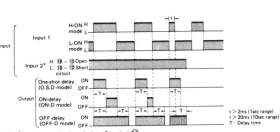

NOTE: *For timer operation, leave the input 2 terminal ⑫ open. When the level of input signal voltage at the input 2 terminal is low or the input 2 terminal is connected to 0V, the output relay releases without regard to the operation of the abovementioned input signal. The input 2 thus connected can be used as an inhibit signal.

- **Type S3S-P10-US**

The power type controller unit for 24 VDC supply performs the same operation as the standard type.

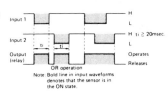

Note. Bold line in input waveforms denotes that the sensor is in the ON state.

■ **DIMENSIONS** [Unit: mm (inch)]

- **CONTROLLER UNIT**

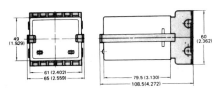

Note: Each of these units is supplied with Type P2A-12B-US connecting socket.

- **CONNECTING SOCKET (Type P2A-12B-US)**

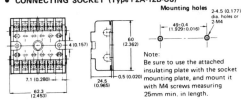

Note:
Be sure to use the attached insulating plate with the socket mounting plate, and mount it with M4 screws measuring 25mm min. in length.

Model S3S — OMRON — Cat. No. Q30-E3-4

■ CONNECTIONS

● S3S-A10-US
Terminals ⑩, ⑪ and ⑫ may also be used to connect another sensor. In this case, note that the controller unit performs AND operation when two sensors are connected to ⑦, ⑧, ⑨ and ⑩, ⑪, ⑫ respectively.

● S3S-B10-US
1. Timer operation
   (1) Use terminals ⑦, ⑧ and ⑨.
   (2) Never use terminal ⑫ as it is used for the memory operation only.
   (3) Terminals ⑩ and ⑪ may be used for DC power supply.
2. Memory operation
   (1) Be sure to set the ON- and OFF- delay variable resistors of the timer, respectively to the "MIN." position.
   (2) When the timer is not set at the "MIN." position, the timer may self-hold upon application of power. However, this should not be considered as trouble.
   (3) When set and reset signals are input simultaneously, the set input signal takes precedence over the reset input signal.

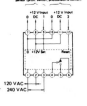

● S3S-C10
1. Timer operation
   (1) Use terminals ⑦, ⑧ and ⑨ for timer operation.
   (2) Never use terminal ⑫.
   (3) Terminals ⑩ and ⑪ may be used for DC power supply.
2. AND operation
   When input 1 and input 2 are applied simultaneously, the input 1 (terminal ⑨) takes precedence over the input 2 (terminal ⑫).

● S3S-P10-US
Terminals ⑩, ⑪ and ⑫ may also be used to connect another sensor. Note that the controller unit performs OR operation when two sensors are connected to ⑦, ⑧, ⑨ and ⑩, ⑪, ⑫ respectively.

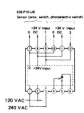

■ APPLICATION EXAMPLES

● Inhibit input operation
Regardless of the operation of the sensor, the output relay releases upon application of the inhibit input signal (by connecting terminals ⑩ and ⑫).

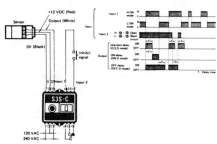

● AND operation
When two sensors are directly connected to the Type S3S-C10 controller unit, the output relays operates only when both sensors are in the ON state (with mode selector set to H-ON).

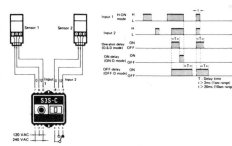

● OR operation
In this example of circuit consisting of two sensors, Type S3S-C10 controller unit and diodes, output turns on when either of the sensors is in the ON state.

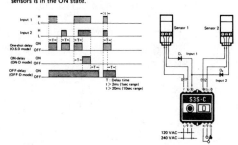

APPLICATION EXAMPLES

- Single-unit control
  - For separate type photoelectric sensors

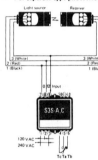

Applicable models:
- E3N and E3S photoelectric sensors

NOTE: Use of S3S-P10-US is recommended for the current output type sensors (rated at 24 VDC).

- For various electric switches

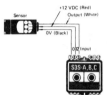

Applicable models:
- TL, and E2K proximity sensors
- E3N and E3S(-L) photoelectric sensors
- E7A solid-state level switches
- Others

NOTES:
1. Use of S3S-P10-US is recommended for the current output type sensors (rated at 24 VDC).
2. Use S3S-B-US for timer operation. In this case, leave terminal ⑫ open.

- Memory operation
  - For external resetting

Even if the sensor turns off, the output relay remains in the operate state until the reset button is turned off. Therefore, this example is ideal for burglar prevention, abnormality alarm, etc.

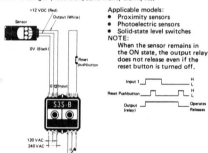

Applicable models:
- Proximity sensors
- Photoelectric sensors
- Solid-state level switches

NOTE: When the sensor remains in the ON state, the output relay does not release even if the reset button is turned off.

- 2-unit control

For level control and positioning control in reciprocating operation

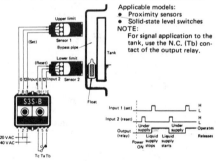

Applicable models:
- Proximity sensors
- Solid-state level switches

NOTE: For signal application to the tank, use the N.C. (Tb) contact of the output relay.

- AND operation

For length detection and displacement identification
In this example of circuit consisting of plural sensors, S3S-A10-US controller unit, resistor and diodes, output turns ON only when each sensor is in the ON state.

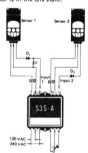

Applicable models:
- Proximity sensors
- Solid-state level switches
- Others

NOTES:
1. R = 10kΩ
   D_1, D_2 = Diodes
2. When the sensors 1 and 2 can be wire ANDed, D_1, D_2 and R are not required.

- OR operation

For extended detecting range
In this example of circuit consisting of plural sensors, S3S-A10-US controller unit and diodes, output turns on when any of the sensors is in the ON state.

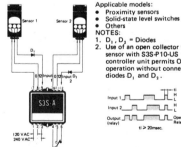

Applicable models:
- Proximity sensors
- Solid-state level switches
- Others

NOTES:
1. D_1, D_2 = Diodes
2. Use of an open collector type sensor with S3S-P10-US controller unit permits OR operation without connecting diodes D_1 and D_2.

INDEX

A

Accuracy, 17
Air pressure grippers, 72
Angular grippers, 65
Applications:
    assembly, 80
    barriers to, 5
    cylindrical coordinate system, 32
    drilling, 77
    first, 4
    grinding, 79
    late start, 5
    non-assembly, 45
    painting, 78
    rectilinear coordinate system, 29
    sealers, 77
    welding, 77
Arm geometries, 8, 25
    cylindrical, 25, 29
    jointed-spherical, 37
    rectangular, 25, 27
    spherical, 25, 34
Arm motion:
    orientation axes, 10, 17, 27, 29, 34, 37
    pitch, 10, 27
    position axes, 10, 17, 27, 29, 34, 37
    roll, 10, 27
    yaw, 10, 27
Artificial skin, 27, 87
Asimov's laws, 187

B

Ballscrew drives, 42
Bang-bang robot, 41
Branching in programs, 53

C

Capek, 3
Cartesian Coordinates, 9, 19, 50, 53
Charge-coupled devices, 143
Charles Stark Draper Laboratories, 83
Closed-loop systems, 45, 100
    advantages, 46, 101
    disadvantages, 47
    feedback sensors, 101
Clustering in vision systems, 145
Compliance, 81
    active, 82
    center of, 82
    lateral, 83
    passive, 82
    remote center compliance (RCC), 82
    rotational, 83
    SCARA robot arm with, 84
Conservation of job I.Q., 63
Contact sensors, 122
Continuous path, 56
    applications, 57
    points stored, 57
    teaching, 57
Control techniques, 45
Controlled path, 55
    straight line motion, 55
    teaching, 55
Cylindrical coordinate system, 29
    advantages, 30
    applications, 32
    disadvantages, 32

D

Definitions:
    accuracy, 17
    compliance, 81
    degree of freedom, 17
    end of arm tooling, 63
    interface, 151
    multiple gripper systems, 85
    orientation axes, 17
    payload, 19
    position axes, 17
    programming robots, 167
    reference frames, 94
    repeatability, 17
    reprogrammable, 7
    robot, 7
    sensors, 121

Definitions (*cont.*)
   servomechanisms, 93
   simple sensors, 152
   speed, 19
   tool-center-point, 18
   work cell coordinates, 19
   work envelope, 17
Degree of freedom, 17, 27, 29, 34
Devol, 4
Dry circuit rating, 123

E

Edge detection in vision, 145
Electronic duty, 123
End effector, 10, 63
End of arm tooling, 10, 61
   air pressure gripper, 72
   angular gripper, 65
   characteristics, 63
   classification, 65
   magnetic grippers, 72
   mandrel gripper, 74
   parallel gripper, 65
   pin gripper, 75
   pneumatic fingers, 73
   power sources, 66
   special purpose grippers, 76
   special purpose tools, 76
   standard grippers, 65
   tactile sensing, 67
   vacuum cup operation, 70
   vacuum cups, 68
   vacuum grippers, 68
   vacuum suckers, 72
   vacuum surfaces, 71

F

Feedback sensors, 93
   encoder, 46, 93
   potentiometer, 46, 93, 101
   resolver, 46, 93
   tachometer, 46
Fixed stops, 41, 47

G

Gantry type robot, 29
Grey scale, 145
Gripper, 10, 63

H

Haptic perception, 129
High technology robots, 57
History, 3
Hohn, 4

I

Interfaces, 151
   complex sensor, 158
   guidelines, 154
   robot control, 155
   simple, 152
   wrist, 154

J

Joint angles, 19
Jointed-spherical coordinate system, 36, 42
   advantages, 38
   disadvantages, 39

L

Language development, 167
Limit, 47
Limit switches, 122
   dog, 123
   electrical properties, 122
   electronic duty, 123
   operational characteristics, 123
   physical properties, 122
   pilot duty, 123
   trip dog design, 125
Limit sensing, 48
Low technology robot, 57

M

Multiple end effector system, 85
   advantages, 85
   disadvantages, 85

N

Non-contact sensors, 129
Non-servo system, 45

Index 229

O

Open-loop systems, 47, 96
    adjustable hard stops, 97
    advantages, 97
    drum-programmers, 98
    fixed hard stops, 97
    limit switches, 98
    operation, 100
    pneumatics, 99
    programmable controllers, 99
    stepper motors, 98
Organized labor, 197
Orientation axes, 17
OSHA, 187

P

Parallel grippers, 65
Path control, 25, 49
Payload, 19
Pendant, 15, 51
Permanent program storage, 14
Photoelectric sensors:
    applications, 141
    applications procedure, 140
    four types, 135, 137
    operating parameters, 136
Pick-and-place robot, 41, 48
Pilot duty, 123
Pixel, 145
Point-to-point control, 51
    arm motion, 54
    comparison to continuous path, 56
    controlled path, 55
    points stored, 53, 56
    teaching, 53, 56
Position axes, 17
Power sources, 8, 10, 39
    cylindrical geometry, 30
    electric, 41
    electric advantages, 42
    electric disadvantages, 42
    hydraulic, 39
    hydraulic advantages, 39
    hydraulic disadvantages, 39
    jointed-spherical geometry, 38
    pneumatic, 41
    pneumatic advantages, 41
    pneumatic disadvantages, 41
    rectangular geometry, 27
    spherical geometry, 34
Process sensor, 149
Production tooling, 10
Program functions, 167

Programming languages:
    AL, 168
    AML, 168
    AUTOPASS, 173
    cycle, 175
    Help, 171
    Joint control, 169
    LAMA, 173
    MCL, 171
    PAL, 171
    primitive motion, 170
    RAPT, 173
    RPL, 171
    SAIL, 168
    Sigla, 170
    structured, 171
    task oriented, 172
    T3, 168, 173
    VAL, 171
Programming levels, 169
Programming offline, 17
Program point storage, 49
Proximity sensors, 129
    application procedure, 134
    detection process, 133
    operating parameters, 131
    package types, 130
    sensing distance variation, 134
Pump and tank, 39

R

Reasons for rapid development, 5
Rectilinear coordinate system, 27
    advantages, 27
    disadvantages, 29
Reference frames, 94
Remote center compliance (RCC), 82
Repeatability, 17
Resistance to robots, 196
Robot arm, 8
Robot controller, 13
Robot Institute of America, 7
Robot systems, 8

S

Safety, 187
Safety guidelines, 190
Safety of operator, 189
Safety of personnel, 188
Scheinman, 4
Selective compliance articulated robot arm (SCARA), 84

Sensors:
  applications of, 121
  contact, 122
  non-contact, 129
  process, 149
  tactile, 127
  vision, 142
Sensors for active compliance, 82
Servo-motor, 41
Servo-system, 45, 93
Simple touch, 127
Social implications, 193
Speed, 19
Special purpose gripper, 76
Special purpose tools, 76
Spherical coordinate system, 34
  advantages, 35
  disadvantages, 35
SRI algorithm, 148
Stepper motor, 41
Stop-to-stop control, 51
Stop-to-stop robot, 48
Straight line motion, 55

T

Tactile sensing, 87
  definition, 127
  haptic perception, 127
  parameters measured, 127
  simple touch, 127
Teach station, 14
  controller front panel, 16
  function, 16
  pendants, 15, 51
  terminals, 15
Template matching, 147
Tool center point, 18
Tool plate, 10
Training activities, 193

Training for maintenance, 195
Training for operators, 195
Training in general, 194
Transformations, 94
Trip dog, 123
T3-566 robot, 11

V

Vacuum gripper, 68
Vacuum sucker, 72
Vacuum surface, 71
Vision, 87
Vision systems:
  CCD camera, 143
  clustering in, 146
  components, 143
  edge and region statistics, 148
  edge detection, 145
  five tasks of, 142
  grey scale, 145
  image analysis, 145
  image measurement, 145
  image recognition, 147
  pixel, 145
  region growing, 146
  resolution, 145
  SRI algorithm, 148
  template matching, 147
  vidicon camera, 144

W

Welding weave patterns, 55
Wiener, 4
Work cell coordinates, 9, 19, 50, 53
Work cell system, 8
Work envelope, 17